Fundamental Mathematics:

Problems and Exercises

Fundamental Mathematics:

Problems and Exercises

G.R. Gilbart-Smith
R.Q. Edmonds
D.J. Hillard

Hodder & Stoughton

A MEMBER OF THE HODDER HEADLINE GROUP

ISBN 0 7131 0855 X

First published 1984
Impression number 16 15 14 13 12 11 10 9 8 7
Year 1999 1998 1997 1996 1995 1994

Typeset by Graphicraft Typesetters Limited, Hong Kong.
Printed in Great Britain for Hodder & Stoughton Educational, a division of Hodder Headline Plc, 338 Euston Road, London NW1 3BH by Bath Press.

Preface

This textbook comprises a two-year course for 11–13 year old pupils. The aim is to provide a firm foundation upon which subsequent work to 'O' level or CSE can be based, though some teachers may choose to use it as a source of revision or remedial material for older pupils. Earlier chapters contain some material normally covered in Primary Schools to give a gentle lead in for the less able. Throughout the book a careful balance has been maintained between 'traditional' and 'modern' topics to ensure its suitability for a wide range of syllabuses, and particular attention has been given to the needs of 11–13 year olds in independent schools who are preparing for the Common Entrance Examination.

Twenty years ago, at the start of the modern mathematics revolution, it was necessary to produce textbooks with a lot of text and notes, as much for the guidance of teachers as of students. This led to the recurring criticism that there were not enough examples to allow sufficient practice. Since then, the more experienced teachers have become adjusted to the details of the course, while many of those new to the profession were themselves brought up on similar courses. Pages of explanation are thus less necessary, and today, with soaring costs of production and slender resources, anything unnecessary is wasteful. So it is the authors' conviction that the modern textbook, especially one aimed at the middle school, best caters for the pupils' needs by providing plenty of examples. Accordingly, this book includes a few brief introductory notes only where this is thought advisable, and devotes the bulk of its space to over 5000 examples. These have been carefully graded to allow a gentle introduction to new material while yet providing work of an imaginative and demanding nature for the more able. Especial care has been taken to ensure that parts of the subject often found difficult by pupils of average and below average ability are given very thorough coverage.

The exercises have been grouped into chapters and the chapters ordered logically, but this is not meant to dictate a teaching order. Broadly speaking, chapters 1 to 17 might be expected to be covered in the first year and chapters 18 to 30 in the second but there will be considerable variation from course to course. Many chapters contain at the end some more advanced material best left for a second 'bite', and some of the larger chapters, notably those covering 'traditional' algebra, are probably best taught a few exercises at a time with frequent shorter visits for practice and revision. It is hoped that Department Heads will have ample scope to draw up detailed syllabuses and that teachers will be able to adapt the speed and depth of cover to suit their individual classes. Above all, it is hoped that nobody will run short of examples!

G.R. Gilbart-Smith
R.Q. Edmonds
D.J. Hillard
April, 1984

Contents

1 The four rules 1
1.1 Addition and subtraction; 1.2 Multiplication; 1.3 Division; 1.4 Problems

2 Introduction to number bases 3
2.1 Conversion to denary scale; 2.2 Conversion from denary

3 Fractions 5
3.1 Equivalent fractions; 3.2 Lowest terms; 3.3 Conversion to whole numbers; 3.4 Conversion to mixed numbers; 3.5 Lowest Common Multiple; 3.6 Addition of fractions; 3.7 Addition of mixed numbers; 3.8 Subtraction of fractions; 3.9 Subtraction of mixed numbers; 3.10 Subtraction from whole numbers; 3.11 Subtraction – borrowing; 3.12 Addition and subtraction; 3.13 Multiplication of proper fractions; 3.14 Conversion to improper fractions; 3.15 Multiplication; 3.16 Division of fractions; 3.17 Division of mixed numbers; 3.18 Multiplication and division; 3.19 Problems

4 Decimals 10
4.1 Place value; 4.2 Conversion to fractions; 4.3 Addition; 4.4 Subtraction; 4.5 Addition and subtraction; 4.6 Multiplication by powers of 10; 4.7 Short multiplication; 4.8 Long multiplication; 4.9 Division by powers of 10; 4.10 Short division; 4.11 Conversion from fractions; 4.12 Ordering

5 Measurement 13
5.1 Money; 5.2 Money problems; 5.3 The metric system; 5.4 Fractions of quantities; 5.5 Proportional parts; 5.6 Four operations involving metric units; 5.7 Problems; 5.8 Time; 5.9 Problems on time

6 Factors 19
6.1 Prime numbers; 6.2 Products of factors; 6.3 Prime factors; 6.4 Prime factors in index form; 6.5 Square roots; 6.6 Highest Common Factor; 6.7 Lowest Common Multiple; 6.8 Lowest Common Multiple – problems

7 Sets and Venn diagrams 22
Introduction – set symbols; 7.1 Listing members; 7.2 The universal set; 7.3 Membership; 7.4 Number of elements; 7.5 Subsets; 7.6 Relationships between two sets; 7.7 Venn diagrams – regions; 7.8 Problems involving Venn diagrams – two sets

8 Perimeter and area 30
8.1 Squares; 8.2 Rectangles; 8.3 Rectangular figures; 8.4 Mixed units

9 Averages 35
9.1 Mean of whole numbers; 9.2 Mean of fractions and decimals; 9.3 Mean of quantities; 9.4 Missing numbers; 9.5 Problems

10 Distance, speed and time 37
10.1 Simple calculations; 10.2 Further calculations; 10.3 Problems

11 Statistical graphs 40
11.1 Column graphs; 11.2 Pie charts

12 Introduction to algebra 45
12.1 Elementary use of letters; 12.2 Further use of letters; 12.3 Letters in equations; 12.4 Directed numbers; 12.5 Rules of signs; 12.6 Substitution – whole numbers; 12.7 Collecting like terms; 12.8 Further equations – negative and fractional solutions; 12.9 Removing brackets; 12.10 Introduction to formulae; 12.11 Two-stage equations; 12.12 Use of indices; 12.13 Common factors and multiples; 12.14 Equations with brackets; 12.15 Inequalities

13 Angle calculations 56
13.1 Numerical calculations; 13.2 Algebraic calculations; 13.3 Parallel lines

14 Construction of triangles 62
14.1 Lines; 14.2 Angles – circumflex notation; 14.3 Angles – three letter notation; 14.4 Naming angles; 14.5 Construction of triangles; 14.6 Mediators; 14.7 Angle bisectors; 14.8 Perpendiculars; 14.9 Triangles – including ruler and compass constructions

15 Polygon facts 67
15.1 Facts of simple polygons; 15.2 Triangle facts; 15.3 Quadrilateral facts

16 Coordinates 69
16.1 Coordinates – first quadrant; 16.2 Coordinates – four quadrants; 16.3 Lines parallel to the axes and to $y = \pm x$

17 Rotation and reflection 77
17.1 Line symmetry; 17.2 Reflection; 17.3 Rotation

18 Further arithmetic 85
18.1 Order of operations with whole numbers; 18.2 Brackets and fractions; 18.3 Combined operations; 18.4 Long division – whole numbers; 18.5 Long division – decimals; 18.6 Combined operations with decimals

19 Percentage 87
19.1 Relationship to fractions and decimals; 19.2 Percentage calculations; 19.3 One quantity as a percentage of another; 19.4 Problems

20 Ratio 90
20.1 Simple ratio; 20.2 Problems; 20.3 Ratios in the form $1 : x$; 20.4 Ratios in the form $x : 1$; 20.5 Equivalent ratios; 20.6 Further problems; 20.7 Proportion; 20.8 Similarity; 20.9 Increasing and decreasing in ratio; 2.10 Finding the ratio of increase or decrease

21 Degree of accuracy and standard form 96
21.1 Nearest whole number; 21.2 Decimal places; 21.3 Significant figures; 21.4 Estimates; 21.5 Harder estimates; 21.6 Conversion from standard form; 21.7 Conversion to standard form; 21.8 Calculations with standard form

22 Arithmetic in bases other than ten 99
22.1 Addition; 22.2 Subtraction; 22.3 Multiplication of single digits; 22.4 Short multiplication; 22.5 Long multiplication; 22.6 Short division; 22.7 Long division in binary

23 Further work with sets 102
23.1 Introduction to Venn diagrams involving three sets; 23.2 Statements, Venn diagrams and set notation

24 Harder areas and volumes 105
24.1 Parallelogram, triangle and trapezium; 24.2 Composite bodies – surface area; 24.3 Volumes of cuboids; 24.4 Miscellaneous problems – mass, capacity and cost

25 Conversion and travel graphs 113
25.1 Conversion graphs; 25.2 Travel graphs

26 Further algebra 118
26.1 Directed numbers – revision; 26.2 Use of letters – revision; 26.3 Simple revision problems; 26.4 Formulation and solution of problems; 26.5 Further problems for formulation and solution; 26.6 Problems involving brackets; 26.7 Equivalent fractions; 26.8 Addition and subtraction of simple algebraic fractions; 26.9 Addition and subtraction of algebraic fractions – numerical denominators; 26.10 Equations and problems with fractions; 26.11 Equations and problems with fractions and brackets; 26.12 Harder inequalities; 26.13 Inequalities involving fractions or decimals; 26.14 Simplification of fractions; 26.15 Further substitution; 26.16 Rearranging formulae

27 Further work with angles, polygons and solids 134
27.1 Further angle calculations; 27.2 Miscellaneous harder angle calculations; 27.3 Construction of quadrilaterals; 27.4 Further work with polygons; 27.5 Problems with polygons; 27.6 Solid figures

28 Scale drawing 149
28.1 Scales; 28.2 Scale drawings; 28.3 Bearings – drawing; 28.4 Bearings – measurement; 28.5 Scale drawings with bearings

29 Loci 153
29.1 Simple loci; 29.2 Further loci; 29.3 Two-dimensional loci

30 Further transformations 156
30.1 Translation; 30.2 Mixed transformations

1

The four rules

Exercise 1.1 *Addition and subtraction*

Add together the following:

1. Two hundred and seventeen, five thousand six hundred and thirty seven, and twenty six.
2. Two thousand and twenty seven, four hundred and six, and seventy four.
3. One thousand and nine, two hundred and seventy four, and three thousand and fifteen.
4. Six thousand two hundred, nine hundred and eleven, and seventeen thousand and sixty three.
5. Twenty nine thousand and six, two thousand and nine, and nine thousand two hundred and six.

Do these subtractions:

6. $1734 - 967$ **7.** $700 - 146$ **8.** $1010 - 428$ **9.** $3070 - 1436$ **10.** $5000 - 378$

11. From 4763 take 1928.

12. Subtract 4686 from 6430.

13. Take 4702 from 16 000.

14. From 9050 subtract 186.

15. By how much is 865 greater than 180?

16. By how much does five thousand exceed two hundred and ninety seven?

Calculate the following:

17. $47 + 86 - 63$

18. $128 + 67 - 124$

19. $72 + 83 - 148$

20. $116 - 88 + 124$

21. $4 + 6 - 7$

22. $4 - 7 + 6$

23. $10 + 9 - 14$

24. $10 - 14 + 9$

25. $149 - 236 + 174$

26. $838 - 1143 + 596$

27. $43 - 82 + 127 - 79$

28. $39 - 74 - 86 + 144$

Exercise 1.2 *Multiplication*

1. Multiply 798 by (**a**) 2 (**b**) 3 (**c**) 4 (**d**) 5 (**e**) 6
2. Multiply 456 by (**a**) 7 (**b**) 8 (**c**) 9 (**d**) 11 (**e**) 12
3. Multiply 802 by (**a**) 7 (**b**) 4 (**c**) 9 (**d**) 3 (**e**) 12

Find the products of:

4. 29 and 27 **5.** 47 and 36 **6.** 28 and 25 **7.** 71 and 28 **8.** 47 and 96

Do these multiplications:

9. 896×32 **10.** 421×36 **11.** 42×629 **12.** 142×49 **13.** 309×54

Find the product of each pair:

14. 38 and 529 **15.** 608 and 47 **16.** 73 and 264 **17.** 86 and 975
18. 2716 and 43 **19.** 1298 and 65 **20.** 59 and 7329

Exercise 1.3 *Division*

Divide:

1. 5378 by 2 **2.** 5430 by 3 **3.** 7172 by 4 **4.** 9325 by 5 **5.** 9336 by 6
6. 9128 by 7 **7.** 8336 by 8 **8.** 3681 by 9 **9.** 1672 by 11 **10.** 10 992 by 12

Do these divisions:

11. $848 \div 16$ **12.** $1708 \div 28$ **13.** $2940 \div 35$ **14.** $912 \div 48$ **15.** $3096 \div 72$

What is the quotient when the first number is divided by the second?

16. 3976, 14 **17.** 14 366, 22 **18.** 51 352, 56 **19.** 19 755, 45 **20.** 46 560, 80

Exercise 1.4 *Problems*

1. George has 3 envelopes containing 147, 119 and 208 stamps.
(**a**) How many stamps does he have altogether?
(**b**) How many does he need to make his total up to a thousand?

2. A man earns £355 each month. How much does he earn in a year?

3. How many days are there in the last six months of the year?

4. A boy spends two hours eating each day. How many hours does he spend eating in a period of six weeks?

5. Sharon buys a sheet of stamps which contains 20 rows with 10 stamps in each row. She uses 64 stamps. How many stamps are left unused?

6. Tulip bulbs are sold in packets of 20. Adam needs 84 bulbs. How many packets must he buy and how many bulbs are not used?

7. (**a**) Sally gets 37 out of 50 and 42 out of 65 for two exercises. What is the total number of marks she could have got?
(**b**) On the same exercises Susan loses a total of 43 marks. Who gets the most marks and by how much?

8. Paul has 35 model cars, Michael has 29 and Nicholas has 17. If they shared them equally among themselves, how many cars would each have?

9. A box contains 24 bars of chocolate. A shopkeeper has 150 bars. How many boxes can he fill and how many bars are left over?

10. What is the remainder when 5420 is divided by 23?

2

Introduction to number bases

Exercise 2.1 *Conversion to denary scale*

Convert the following numbers to denary (base 10):

1. 11_2	**2.** 101_2	**3.** 1010_2	**4.** $10\,111_2$	**5.** $110\,111_2$
6. $1\,101\,010_2$	**7.** $1001\,101_2$	**8.** 21_3	**9.** 22_4	**10.** 32_5
11. 52_6	**12.** 43_7	**13.** 37_8	**14.** 34_9	**15.** 121_3
16. 231_4	**17.** 242_5	**18.** 424_6	**19.** 313_7	**20.** 123_8
21. 438_9	**22.** 133_4	**23.** 232_8	**24.** 423_5	**25.** 227_9
26. 1011_2	**27.** 141_6	**28.** 220_3	**29.** 125_7	**30.** 1122_3
31. 1203_4	**32.** 1111_5	**33.** $101\,111_2$	**34.** 2022_3	**35.** $10\,000\,111_2$

Exercise 2.2 *Conversion from denary*

Convert the following denary numbers to binary (base 2):

1. 3	**2.** 5	**3.** 14	**4.** 17	**5.** 26
6. 40	**7.** 47	**8.** 64	**9.** 70	**10.** 85

Convert to base 3:

11. 4 **12.** 15 **13.** 28

Convert to base 4:

14. 7 **15.** 18 **16.** 32

Convert to base 5:

17. 23 **18.** 41 **19.** 67

Convert to base 6:

20. 10 **21.** 27 **22.** 62

Convert to base 7:

23. 12 **24.** 35 **25.** 50

Convert to base 8 (octal):

26. 45 **27.** 75 **28.** 100

Convert to base 9:

29. 27 **30.** 66 **31.** 90

Convert the following denary numbers to the base indicated:

32. 29 to base 2
33. 16 to base 3
34. 14 to base 4
35. 43 to base 2
36. 19 to base 3
37. 59 to base 2
38. 46 to base 3
39. 38 to base 5
40. 125 to base 9
41. 108 to base 8
42. 47 to base 6
43. 83 to base 4
44. 45 to base 7
45. 70 to base 2
46. 66 to base 2

3

Fractions

Exercise 3.1 *Equivalent fractions*

Copy and complete the following:

1. $\frac{1}{2} = \frac{\quad}{4}$ **2.** $\frac{2}{3} = \frac{\quad}{18}$ **3.** $\frac{3}{8} = \frac{\quad}{24}$ **4.** $\frac{5}{6} = \frac{\quad}{30}$ **5.** $\frac{7}{15} = \frac{\quad}{30}$

6. $\frac{1}{2} = \frac{4}{\quad}$ **7.** $\frac{3}{10} = \frac{27}{\quad}$ **8.** $\frac{5}{11} = \frac{35}{\quad}$ **9.** $\frac{11}{12} = \frac{44}{\quad}$ **10.** $\frac{19}{20} = \frac{38}{\quad}$

11. $\frac{6}{12} = \frac{\quad}{2}$ **12.** $\frac{12}{14} = \frac{\quad}{7}$ **13.** $\frac{12}{28} = \frac{\quad}{7}$ **14.** $\frac{35}{100} = \frac{\quad}{20}$ **15.** $\frac{18}{32} = \frac{\quad}{16}$

16. $\frac{16}{28} = \frac{4}{\quad}$ **17.** $\frac{21}{27} = \frac{7}{\quad}$ **18.** $\frac{42}{80} = \frac{21}{\quad}$ **19.** $\frac{16}{100} = \frac{4}{\quad}$ **20.** $\frac{44}{72} = \frac{11}{\quad}$

21. $\frac{2}{5} = \frac{\quad}{10} = \frac{\quad}{20}$ **22.** $\frac{3}{7} = \frac{6}{\quad} = \frac{15}{\quad}$ **23.** $\frac{5}{8} = \frac{15}{\quad} = \frac{\quad}{64}$ **24.** $\frac{5}{9} = \frac{\quad}{18} = \frac{25}{\quad}$ **25.** $\frac{1}{\quad} = \frac{3}{12} = \frac{9}{\quad}$

26. $\frac{20}{40} = \frac{5}{\quad} = \frac{\quad}{2}$ **27.** $\frac{10}{\quad} = \frac{5}{7} = \frac{\quad}{49}$ **28.** $\frac{50}{\quad} = \frac{\quad}{3} = \frac{10}{15}$ **29.** $\frac{42}{72} = \frac{\quad}{12} = \frac{84}{\quad}$ **30.** $\frac{6}{18} = \frac{5}{\quad} = \frac{\quad}{90}$

Exercise 3.2 *Lowest terms*

Write these as fractions in lowest terms:

1. $\frac{3}{6}$ **2.** $\frac{4}{16}$ **3.** $\frac{6}{8}$ **4.** $\frac{10}{15}$ **5.** $\frac{10}{12}$

6. $\frac{14}{18}$ **7.** $\frac{21}{30}$ **8.** $\frac{24}{32}$ **9.** $\frac{16}{20}$ **10.** $\frac{9}{15}$

11. $\frac{24}{30}$ **12.** $\frac{27}{63}$ **13.** $\frac{18}{48}$ **14.** $\frac{35}{70}$ **15.** $\frac{24}{96}$

16. $\frac{24}{60}$ **17.** $\frac{21}{63}$ **18.** $\frac{16}{48}$ **19.** $\frac{15}{100}$ **20.** $\frac{84}{100}$

Exercise 3.3 *Conversion to whole numbers*

Write these as whole numbers:

1. $\frac{2}{2}$ **2.** $\frac{9}{3}$ **3.** $\frac{16}{4}$ **4.** $\frac{42}{6}$ **5.** $\frac{15}{5}$

6. $\frac{18}{2}$ **7.** $\frac{24}{6}$ **8.** $\frac{35}{7}$ **9.** $\frac{100}{25}$ **10.** $\frac{100}{20}$

Exercise 3.4 *Conversion to mixed numbers*

Write these as mixed numbers:

1. $\frac{3}{2}$ **2.** $\frac{7}{4}$ **3.** $\frac{12}{5}$ **4.** $\frac{17}{4}$ **5.** $\frac{23}{3}$

6. $\frac{24}{7}$ **7.** $\frac{35}{6}$ **8.** $\frac{32}{5}$ **9.** $\frac{21}{8}$ **10.** $\frac{14}{3}$

11. $\frac{63}{10}$ **12.** $\frac{43}{8}$ **13.** $\frac{34}{9}$ **14.** $\frac{37}{15}$ **15.** $\frac{73}{20}$

16. $\frac{63}{8}$ **17.** $\frac{50}{11}$ **18.** $\frac{29}{13}$ **19.** $\frac{54}{25}$ **20.** $\frac{29}{12}$

Exercise 3.5 *Lowest Common Multiple*

Find the Lowest Common Multiples of the following numbers:

1. 2, 3 **2.** 4, 5 **3.** 2, 7 **4.** 10, 5 **5.** 3, 9

6. 3, 4 **7.** 4, 7 **8.** 3, 6 **9.** 5, 2 **10.** 2, 9
11. 15, 5 **12.** 12, 8 **13.** 9, 12 **14.** 7, 5 **15.** 8, 4
16. 4, 12 **17.** 2, 4, 6 **18.** 5, 10, 15 **19.** 2, 5, 10 **20.** 3, 9, 6
21. 4, 7, 2 **22.** 2, 3, 5 **23.** 4, 5, 8 **24.** 4, 8, 2 **25.** 3, 4, 12

Exercise 3.6 *Addition of fractions*

Add the following fractions, leaving your answers as proper fractions or mixed numbers:

1. $\frac{1}{2}+\frac{1}{3}$ **2.** $\frac{1}{4}+\frac{2}{5}$ **3.** $\frac{3}{10}+\frac{2}{5}$ **4.** $\frac{1}{3}+\frac{4}{9}$ **5.** $\frac{2}{3}+\frac{1}{4}$
6. $\frac{2}{3}+\frac{1}{6}$ **7.** $\frac{1}{2}+\frac{1}{4}+\frac{1}{6}$ **8.** $\frac{1}{5}+\frac{3}{10}+\frac{2}{15}$ **9.** $\frac{3}{5}+\frac{1}{2}$ **10.** $\frac{9}{10}+\frac{4}{5}$
11. $\frac{6}{7}+\frac{1}{2}$ **12.** $\frac{3}{4}+\frac{2}{3}$ **13.** $\frac{7}{15}+\frac{4}{5}$ **14.** $\frac{5}{8}+\frac{7}{12}$ **15.** $\frac{5}{12}+\frac{7}{9}$
16. $\frac{5}{7}+\frac{3}{5}$ **17.** $\frac{7}{8}+\frac{3}{4}$ **18.** $\frac{4}{5}+\frac{1}{2}+\frac{7}{10}$ **19.** $\frac{2}{3}+\frac{7}{9}+\frac{1}{6}$ **20.** $\frac{1}{4}+\frac{6}{7}+\frac{1}{2}$
21. $\frac{1}{2}+\frac{2}{3}+\frac{4}{5}$ **22.** $\frac{3}{4}+\frac{4}{5}+\frac{5}{8}$ **23.** $\frac{1}{6}+\frac{3}{4}+\frac{1}{3}$ **24.** $\frac{1}{4}+\frac{7}{8}+\frac{1}{2}$ **25.** $\frac{2}{3}+\frac{1}{4}+\frac{7}{12}$

Exercise 3.7 *Addition of mixed numbers*

Add, leaving your answers as mixed numbers:

1. $2\frac{1}{4}+1\frac{2}{3}$ **2.** $1\frac{1}{4}+2\frac{3}{5}$ **3.** $3\frac{3}{4}+2\frac{1}{6}$ **4.** $1\frac{7}{8}+2\frac{3}{4}$ **5.** $6\frac{2}{3}+\frac{7}{9}$
6. $5\frac{1}{3}+3\frac{5}{8}$ **7.** $2\frac{3}{7}+1\frac{2}{3}$ **8.** $6\frac{11}{12}+4\frac{1}{3}$ **9.** $5\frac{3}{4}+2\frac{5}{6}$
10. $1\frac{7}{10}+3\frac{11}{15}$ **11.** $3\frac{1}{4}+2\frac{2}{3}+3\frac{1}{6}$ **12.** $4\frac{3}{4}+\frac{7}{16}+3\frac{1}{2}$ **13.** $1\frac{5}{6}+2\frac{1}{3}+3\frac{11}{12}$
14. $2\frac{1}{2}+3\frac{5}{8}+\frac{3}{4}$ **15.** $1\frac{1}{5}+2\frac{3}{4}+1\frac{7}{10}$ **16.** $3\frac{3}{10}+4\frac{1}{6}+1\frac{3}{5}$ **17.** $2\frac{1}{2}+3\frac{2}{3}+1\frac{1}{4}$
18. $2\frac{7}{8}+4\frac{15}{16}+\frac{1}{4}$ **19.** $3\frac{2}{9}+1\frac{1}{2}+4\frac{5}{6}$ **20.** $6\frac{3}{7}+2\frac{1}{2}+3\frac{1}{4}$ **21.** $2\frac{3}{4}+\frac{7}{8}+3\frac{1}{2}$
22. $1\frac{1}{4}+2\frac{3}{5}+4\frac{5}{8}$ **23.** $3\frac{2}{3}+1\frac{5}{6}+2\frac{7}{10}$ **24.** $4\frac{11}{12}+2\frac{5}{8}+3\frac{1}{3}$ **25.** $2\frac{11}{15}+1\frac{5}{12}+3\frac{1}{4}$

Exercise 3.8 *Subtraction of fractions*

Do these subtractions, leaving your answers as proper fractions:

1. $\frac{3}{4}-\frac{1}{2}$ **2.** $\frac{2}{3}-\frac{1}{4}$ **3.** $\frac{7}{16}-\frac{3}{8}$ **4.** $\frac{11}{15}-\frac{3}{5}$ **5.** $\frac{7}{12}-\frac{1}{2}$
6. $\frac{9}{10}-\frac{3}{4}$ **7.** $\frac{11}{12}-\frac{3}{8}$ **8.** $\frac{7}{9}-\frac{1}{6}$ **9.** $\frac{7}{9}-\frac{5}{12}$ **10.** $\frac{5}{6}-\frac{1}{8}$
11. $\frac{3}{4}-\frac{3}{7}$ **12.** $\frac{7}{11}-\frac{1}{3}$ **13.** $\frac{5}{12}-\frac{3}{8}$ **14.** $\frac{7}{9}-\frac{3}{5}$ **15.** $\frac{11}{15}-\frac{7}{12}$

Exercise 3.9 *Subtraction of mixed numbers*

Do these subtractions, leaving your answers as proper fractions or mixed numbers:

1. $2\frac{2}{3}-1\frac{1}{4}$ **2.** $7\frac{3}{4}-4\frac{2}{3}$ **3.** $4\frac{7}{10}-2\frac{1}{5}$ **4.** $3\frac{4}{9}-1\frac{1}{3}$ **5.** $6\frac{4}{5}-3\frac{2}{3}$
6. $7\frac{7}{8}-\frac{1}{4}$ **7.** $3\frac{11}{12}-1\frac{5}{8}$ **8.** $6\frac{3}{5}-1\frac{1}{4}$ **9.** $4\frac{6}{7}-2\frac{2}{3}$ **10.** $5\frac{3}{7}-3\frac{1}{4}$
11. $1\frac{2}{3}-1\frac{3}{10}$ **12.** $4\frac{5}{9}-1\frac{2}{5}$ **13.** $1\frac{7}{8}-\frac{3}{5}$ **14.** $4\frac{3}{10}-2\frac{1}{5}$ **15.** $3\frac{5}{12}-2\frac{1}{15}$

Exercise 3.10 *Subtraction from whole numbers*

Do these subtractions, leaving your answers as proper fractions or mixed numbers:

1. $1-\frac{3}{8}$ **2.** $1-\frac{5}{12}$ **3.** $1-\frac{6}{7}$ **4.** $1-\frac{2}{11}$ **5.** $1-\frac{7}{15}$

6. $2-\frac{1}{2}$	**7.** $6-\frac{3}{4}$	**8.** $3-\frac{5}{16}$	**9.** $2-1\frac{3}{4}$	**10.** $7-5\frac{1}{2}$
11. $3-1\frac{6}{11}$	**12.** $6-3\frac{9}{20}$	**13.** $4-1\frac{7}{10}$	**14.** $8-7\frac{8}{15}$	**15.** $9-1\frac{21}{25}$

Exercise 3.11 *Subtraction – borrowing*

Do these subtractions, leaving your answers as proper fractions or mixed numbers:

1. $6\frac{1}{3}-2\frac{1}{2}$	**2.** $6\frac{1}{2}-1\frac{3}{4}$	**3.** $7\frac{1}{4}-1\frac{2}{3}$	**4.** $4\frac{2}{5}-2\frac{7}{10}$	**5.** $7\frac{1}{6}-4\frac{1}{4}$
6. $6\frac{3}{8}-1\frac{7}{12}$	**7.** $5\frac{1}{2}-4\frac{7}{10}$	**8.** $1\frac{1}{2}-\frac{3}{5}$	**9.** $3\frac{2}{7}-1\frac{1}{2}$	**10.** $7\frac{3}{8}-4\frac{2}{3}$
11. $2\frac{1}{12}-1\frac{3}{4}$	**12.** $1\frac{1}{4}-\frac{7}{8}$	**13.** $7\frac{3}{10}-2\frac{7}{15}$	**14.** $4\frac{1}{6}-1\frac{2}{3}$	**15.** $1\frac{2}{3}-\frac{13}{15}$

Exercise 3.12 *Addition and subtraction*

Work out the following. Leave your answers as whole numbers, proper fractions or mixed numbers:

1. $\frac{1}{4}+\frac{1}{2}-\frac{1}{3}$	**2.** $\frac{2}{5}-\frac{1}{2}+\frac{7}{10}$	**3.** $\frac{11}{12}-\frac{1}{4}-\frac{2}{3}$	**4.** $\frac{11}{15}-\frac{9}{10}+\frac{2}{3}$
5. $\frac{5}{6}+\frac{1}{2}-\frac{2}{5}$	**6.** $\frac{4}{9}+\frac{2}{3}-\frac{5}{6}$	**7.** $3\frac{1}{2}+2\frac{3}{4}-4\frac{2}{3}$	**8.** $3\frac{3}{10}-1\frac{1}{2}+2\frac{4}{5}$
9. $5\frac{15}{16}-1\frac{5}{8}-2\frac{1}{4}$	**10.** $1\frac{3}{5}-4\frac{3}{4}+3\frac{1}{2}$	**11.** $7\frac{9}{10}-3\frac{1}{5}-2\frac{1}{2}$	**12.** $3\frac{5}{6}-2\frac{1}{2}+4\frac{4}{5}$
13. $3\frac{7}{9}-1\frac{1}{4}-1\frac{1}{3}$	**14.** $3\frac{1}{2}-1\frac{2}{3}-1\frac{1}{4}$	**15.** $4\frac{3}{4}-1\frac{5}{6}-2\frac{1}{3}$	**16.** $4\frac{2}{5}+2\frac{1}{10}-3\frac{3}{4}$
17. $1\frac{1}{3}-2\frac{5}{6}+3\frac{1}{4}$	**18.** $5\frac{3}{8}-1\frac{7}{12}-2\frac{1}{4}$	**19.** $1\frac{11}{24}-4\frac{7}{8}+5\frac{1}{6}$	**20.** $4\frac{1}{5}+2\frac{3}{10}-3\frac{1}{2}$

Exercise 3.13 *Multiplication of proper fractions*

Do these multiplications, leaving your answers as proper fractions in lowest terms:

1. $\frac{3}{16}\times\frac{4}{9}$	**2.** $\frac{8}{33}\times\frac{11}{12}$	**3.** $\frac{9}{16}\times\frac{7}{18}$	**4.** $\frac{14}{15}\times\frac{10}{21}$
5. $\frac{9}{14}\times\frac{7}{15}$	**6.** $\frac{8}{9}\times\frac{15}{16}$	**7.** $\frac{5}{11}\times\frac{22}{25}$	**8.** $\frac{9}{20}\times\frac{10}{27}$
9. $\frac{17}{35}\times\frac{15}{34}$	**10.** $\frac{22}{25}\times\frac{15}{44}$	**11.** $(\frac{3}{4})^2$	**12.** $(\frac{1}{2})^4$
13. $(\frac{1}{4})^3$	**14.** $\frac{1}{2}$ of $\frac{4}{5}$	**15.** $\frac{4}{5}$ of $\frac{15}{22}$	**16.** $\frac{1}{3}$ of $\frac{1}{9}$
17. $\frac{2}{3}\times\frac{3}{5}\times\frac{5}{6}$	**18.** $\frac{5}{6}\times\frac{7}{12}\times\frac{9}{14}$	**19.** $\frac{7}{10}\times\frac{11}{14}\times\frac{25}{33}$	**20.** $\frac{9}{16}\times\frac{8}{21}\times\frac{14}{15}$
21. $\frac{3}{4}\times\frac{5}{8}\times\frac{4}{9}$	**22.** $\frac{5}{16}\times\frac{8}{25}\times\frac{4}{15}$	**23.** $\frac{21}{22}\times\frac{5}{7}\times\frac{11}{15}$	**24.** $\frac{9}{26}\times\frac{13}{15}\times\frac{5}{6}$
25. $\frac{25}{28}\times\frac{7}{10}\times\frac{8}{35}$			

Exercise 3.14 *Conversion to improper fractions*

Write as improper fractions:

1. $1\frac{1}{2}$	**2.** $1\frac{3}{4}$	**3.** $2\frac{1}{2}$	**4.** $2\frac{3}{4}$	**5.** $3\frac{1}{3}$
6. $1\frac{7}{12}$	**7.** $2\frac{4}{5}$	**8.** $3\frac{5}{6}$	**9.** $2\frac{7}{10}$	**10.** $3\frac{4}{15}$
11. $3\frac{3}{20}$	**12.** $2\frac{5}{9}$	**13.** $1\frac{13}{22}$	**14.** $3\frac{7}{8}$	**15.** $8\frac{3}{4}$

Exercise 3.15 *Multiplication*

Multiply the following. Do not leave your answers as improper fractions.

1. $3\frac{1}{3}\times1\frac{2}{5}$	**2.** $1\frac{1}{5}\times2\frac{2}{9}$	**3.** $2\frac{2}{5}\times1\frac{1}{8}$	**4.** $1\frac{4}{7}\times1\frac{13}{22}$

5. $1\frac{1}{9} \times 3\frac{3}{5}$
6. $7\frac{1}{2} \times 1\frac{1}{10}$
7. $6 \times 1\frac{1}{9}$
8. $1\frac{7}{8} \times 2\frac{4}{5}$
9. $\frac{9}{14}$ of $1\frac{1}{6}$
10. $2\frac{1}{4} \times \frac{10}{27}$
11. $(2\frac{1}{2})^2$
12. $(1\frac{1}{2})^3$
13. $2\frac{1}{4} \times 1\frac{1}{2} \times 3\frac{1}{3}$
14. $1\frac{3}{5} \times 2\frac{1}{2} \times 1\frac{3}{4}$
15. $1\frac{1}{7} \times 2\frac{2}{5} \times 1\frac{2}{3}$
16. $2\frac{1}{2} \times 1\frac{3}{4} \times \frac{8}{15}$
17. $\frac{2}{5} \times 4\frac{1}{2} \times 2\frac{2}{3}$
18. $1\frac{5}{16} \times (1\frac{1}{3})^2$
19. $\frac{3}{4}$ of $(1\frac{1}{2})^2$
20. $4\frac{1}{2} \times 2\frac{1}{4} \times 1\frac{1}{3}$

Exercise 3.16 *Division of fractions*

Divide the following. Do not leave your answers as improper fractions.

1. $3 \div \frac{1}{2}$
2. $4 \div \frac{1}{4}$
3. $4 \div \frac{1}{3}$
4. $6 \div \frac{1}{8}$
5. $\frac{1}{2} \div \frac{3}{4}$
6. $\frac{2}{3} \div \frac{5}{6}$
7. $\frac{5}{6} \div \frac{11}{12}$
8. $\frac{5}{8} \div \frac{15}{16}$
9. $\frac{3}{5} \div \frac{9}{10}$
10. $\frac{7}{8} \div \frac{3}{16}$
11. $\frac{3}{4} \div \frac{1}{2}$
12. $\frac{11}{12} \div \frac{5}{6}$
13. $\frac{9}{11} \div \frac{3}{22}$
14. $\frac{7}{15} \div \frac{2}{3}$
15. $\frac{5}{16} \div \frac{5}{8}$
16. $\frac{17}{21} \div \frac{2}{3}$
17. $10 \div \frac{4}{5}$
18. $4 \div \frac{2}{3}$
19. $5 \div \frac{3}{4}$
20. $\frac{2}{3} \div 3$
21. $\frac{4}{7} \div 4$
22. $\frac{1}{2} \div 2$
23. $\dfrac{\frac{2}{3}}{\frac{1}{2}}$
24. $\dfrac{\frac{1}{6}}{\frac{3}{4}}$
25. $\dfrac{\frac{11}{15}}{\frac{3}{5}}$

Exercise 3.17 *Division of mixed numbers*

Divide the following. Do not leave your answers as improper fractions.

1. $2\frac{1}{2} \div 3\frac{1}{3}$
2. $1\frac{2}{7} \div 1\frac{1}{14}$
3. $5\frac{1}{2} \div 4\frac{1}{8}$
4. $3\frac{2}{5} \div 2\frac{4}{15}$
5. $3 \div \frac{2}{3}$
6. $7\frac{1}{2} \div \frac{5}{8}$
7. $\frac{5}{8} \div 1\frac{1}{4}$
8. $1\frac{1}{21} \div 2\frac{2}{7}$
9. $4\frac{1}{2} \div \frac{3}{10}$
10. $1\frac{3}{4} \div 2\frac{5}{8}$
11. $\frac{3}{4} \div 6$
12. $1\frac{7}{8} \div \frac{5}{24}$
13. $9\frac{1}{3} \div 1\frac{1}{6}$
14. $12 \div 1\frac{1}{5}$
15. $8\frac{3}{4} \div 3\frac{1}{2}$
16. $1\frac{7}{9} \div 3\frac{1}{3}$
17. $2\frac{3}{4} \div 1\frac{1}{6}$
18. $\dfrac{1\frac{3}{4}}{2\frac{5}{8}}$
19. $\dfrac{2\frac{2}{3}}{1\frac{1}{15}}$
20. $\dfrac{\frac{3}{4}}{8}$

Exercise 3.18 *Multiplication and division*

Work out the following. Do not leave your answers as improper fractions.

1. $\frac{3}{4} \times \frac{8}{9} \div \frac{5}{6}$
2. $\frac{7}{8} \div \frac{5}{12} \times \frac{10}{21}$
3. $\frac{3}{4}$ of $\frac{14}{15} \div \frac{3}{10}$
4. $\frac{11}{16} \times \frac{9}{22} \div \frac{3}{8}$
5. $\frac{9}{10} \div \frac{4}{5} \times \frac{16}{27}$
6. $\frac{1}{2} \times \frac{3}{8} \div \frac{9}{16}$
7. $\frac{20}{21} \div \frac{2}{7} \div \frac{2}{3}$
8. $6\frac{1}{4} \times 1\frac{1}{5} \div 2\frac{1}{4}$
9. $1\frac{5}{6} \div 2\frac{1}{2} \times 1\frac{7}{8}$
10. $7\frac{1}{2} \div 1\frac{2}{3} \times 1\frac{1}{9}$
11. $2\frac{1}{2} \div 2\frac{2}{3} \div 3\frac{3}{4}$
12. $7\frac{1}{2} \times 1\frac{3}{5} \div 2\frac{1}{2}$
13. $4\frac{1}{2} \div 2\frac{2}{3} \times 3\frac{1}{3}$
14. $(\frac{3}{4})^2 \times 5\frac{1}{3} \div 1\frac{7}{8}$
15. $(1\frac{1}{2})^2 \div 4\frac{1}{2} \times 2\frac{2}{5}$

Exercise 3.19 *Problems*

1. What is the sum of three quarters and four fifths?
2. What is the difference between $2\frac{1}{3}$ and $1\frac{1}{2}$?
3. What must be added to seven eighths to make $2\frac{1}{4}$?
4. By how much does $3\frac{7}{10}$ exceed $1\frac{7}{8}$?
5. What is the product of $7\frac{1}{2}$ and $2\frac{2}{3}$?
6. How many quarters are there in $5\frac{1}{2}$?
7. How many times is three quarters contained in six?
8. By what must $7\frac{1}{3}$ be multiplied to give an answer of two?
9. What number is $1\frac{1}{2}$ times greater than $1\frac{1}{3}$?

10. What number is greater than $1\frac{1}{3}$ by $1\frac{1}{2}$?

11. $\frac{9}{20}$ of a school of 600 pupils are girls. How many boys are there?

12. A man earns £100 per week.

(**a**) He spends £35 on food. What fraction of his income is this? Give your answer in lowest terms.

(**b**) He saves $\frac{3}{20}$ of his weekly income. How much does he save in 4 weeks?

13. There are 18 girls and 14 boys in a class. Altogether 20 pupils play a musical instrument. If $\frac{4}{9}$ of the girls play, what fraction of the boys play an instrument? Give your answer in lowest terms.

14. $\frac{5}{6}$ of the houses in a road have a television. There are 150 houses in the road. How many have not got a television?

15. The number of pupils in a school has increased by $\frac{3}{20}$. If there were 580 pupils, how many are there after the increase?

16. John gets 65 on a Maths paper. Tony gets $\frac{10}{13}$ of John's total. What is Tony's mark?

17. A coach starts its journey with 48 passengers. $\frac{1}{3}$ of these get off at the first stop and none get on. $\frac{3}{4}$ of the remainder get off at the next stop. How many are left on the coach?

18. 200 buns are delivered to the baker. $\frac{1}{10}$ of these are put on one side for Mrs Adams. $\frac{19}{20}$ of the remainder are sold to customers. How many buns are unsold?

19. A man works from 8.00 a.m. until 5.00 p.m. He spends 1 hour at lunch and has two breaks of 15 minutes each. For what fraction of the working day is he not working?

20. $\frac{3}{16}$ of a journey of 80 miles is on a motorway.

(**a**) How many miles of the journey are not on the motorway?

(**b**) The driver averages 70 m.p.h. on the motorway but averages a speed $2\frac{1}{2}$ times slower on the ordinary road. What is this speed?

4

Decimals

Exercise 4.1 *Place value*

Give the value of each of the underlined digits:

1. 4$\underline{3}$7 **2.** $\underline{2}$008 **3.** 24 $\underline{1}$76 **4.** 39 74$\underline{2}$ **5.** $\underline{2}$8 650
6. 6.$\underline{4}$3 **7.** 7.2$\underline{8}$ **8.** 1.93$\underline{5}$ **9.** $\underline{6}$.264 **10.** 1.$\underline{3}$95
11. 17.6$\underline{3}$2 **12.** 1.94$\underline{3}$ **13.** 180.$\underline{7}$35 **14.** 9.672$\underline{8}$ **15.** 140.8$\underline{6}$

Exercise 4.2 *Conversion to fractions*

Write these as fractions in lowest terms:

1. 0.7 **2.** 0.8 **3.** 0.5 **4.** 0.2 **5.** 0.27
6. 0.35 **7.** 0.36 **8.** 0.25 **9.** 0.72 **10.** 0.85
11. 0.06 **12.** 0.125 **13.** 0.225 **14.** 0.875 **15.** 0.4
16. 0.04 **17.** 0.44 **18.** 0.45 **19.** 0.46 **20.** 0.01

Write as mixed numbers with fractions in lowest terms:

21. 3.6 **22.** 6.75 **23.** 2.375 **24.** 7.24 **25.** 17.625

Exercise 4.3 *Addition*

Add the following:

1. 4.73, 8.98 **2.** 18.43, 9.69 **3.** 23.42, 8.7
4. 9.87, 14.6 **5.** 14.4, 0.375 **6.** 42.7, 3.95, 28
7. 0.436, 27.7, 1.48 **8.** 19.73, 4.635, 28.4 **9.** 28.47, 1.395, 206.8
10. 4.376, 19.4, 0.8 **11.** 19.1, 3.725, 81.92 **12.** 0.47, 38, 4.9
13. 43.65, 1.9, 0.786 **14.** 275, 0.275, 27.5 **15.** 0.65, 6.5, 65

Exercise 4.4 *Subtraction*

Do these subtractions:

1. 18.7 − 12.4 **2.** 3.86 − 1.95 **3.** 7.32 − 4.96 **4.** 21.4 − 8.7
5. 421.6 − 48.8 **6.** 4.38 − 2.1 **7.** 9.48 − 5.6 **8.** 27.35 − 1.8
9. 6.2 − 1.15 **10.** 9.3 − 2.67 **11.** 14.3 − 8.627 **12.** 43 − 4.3
13. Take 1.97 from 8.3
14. From 2.9 subtract 0.84
15. From 43.45 take 9.5
16. Subtract 14.2 from 20
17. What is the difference between 28.7 and 2.87?
18. What must be added to 13.87 to make 50?
19. By how much does 1.75 exceed 0.175?
20. What is the difference between 0.7 and 0.07?

Exercise 4.5 *Addition and subtraction*

Work out the following:

1.	$4.6 + 8.9 - 1.7$	**2.**	$2.71 + 4.63 - 5.86$	**3.**	$3.75 + 2.91 - 4.8$
4.	$7.4 + 8.9 - 11.64$	**5.**	$28.7 + 17.3 - 14.35$	**6.**	$1.63 - 8.7 + 9.8$
7.	$14.3 - 27.8 + 16.926$	**8.**	$7.85 - 9.6 + 3.7$	**9.**	$0.4 - 0.44 + 0.04$
10.	$4.7 - 9.6 + 11.5 - 1.88$				

Exercise 4.6 *Multiplication by powers of 10*

Multiply the following numbers by **(a)** 10 **(b)** 100 **(c)** 1000:

1.	4.637	**2.**	8.64	**3.**	0.5	**4.**	14.62	**5.**	7.4
6.	132.64	**7.**	0.8273	**8.**	27.6	**9.**	12.95	**10.**	0.0376

Multiply the following numbers by **(a)** 10 **(b)** 10^2 **(c)** 10^3:

11.	27.32	**12.**	0.4163	**13.**	2.9	**14.**	0.6	**15.**	1.375
16.	0.0475	**17.**	43.6	**18.**	243.25	**19.**	32.7	**20.**	1.43

Exercise 4.7 *Short multiplication*

Multiply the following:

1.	8.7 by 0.4	**2.**	23.9 by 0.5	**3.**	2.04 by 0.9	**4.**	48.9 by 0.03
5.	1.14 by 0.8	**6.**	2.73 by 0.07	**7.**	1.25 by 0.03	**8.**	1.95 by 0.004
9.	0.47 by 0.9	**10.**	1.32 by 0.0006	**11.**	14.15 by 0.004	**12.**	0.73 by 6

Calculate:

13.	$(0.2)^2$	**14.**	$(0.4)^2$	**15.**	$(0.3)^2$	**16.**	$(0.02)^3$
17.	$(0.5)^3$	**18.**	$(0.1)^4$	**19.**	$(1.1)^2$	**20.**	$(0.12)^2$

Exercise 4.8 *Long multiplication*

Do these multiplications:

1.	14.6×2.3	**2.**	8.75×0.46	**3.**	0.183×2.6	**4.**	8.4×0.37
5.	0.35×0.35	**6.**	4.37×0.46	**7.**	3.78×0.062	**8.**	0.143×0.73
9.	0.0162×1.4	**10.**	37×0.293				

11. **(a)** Work out 27.6×4.2
What is the answer to **(b)** 0.276×42 **(c)** 2.76×4.2?

12. **(a)** Multiply 1.89 by 0.53
Write down the answer to **(b)** 189×5.3 **(c)** 18.9×53

13. **(a)** What is the product of 0.475 and 2.8?
Hence find the answer to **(b)** 47.5×28 **(c)** 0.475×0.28

14. **(a)** Multiply 0.375 by 1.8
Without further working, write down the value of **(b)** 375×18
(c) 0.375×0.018

15. **(a)** Find the product of 28.4 and 7.9
Hence give the value of **(b)** 284×0.079 **(c)** 284×790

Exercise 4.9 *Division by powers of 10*

Divide the following numbers by **(a)** 10 **(b)** 100 **(c)** 1000:

1. 473	**2.** 27.4	**3.** 2763	**4.** 4.9	**5.** 0.725
6. 396	**7.** 3872	**8.** 240	**9.** 6.3	**10.** 49.5

Multiply the following numbers by **(a)** $\frac{1}{10}$ **(b)** $\frac{1}{100}$ **(c)** $\frac{1}{1000}$:

11. 269	**12.** 4.75	**13.** 83.6	**14.** 0.42	**15.** 1927

Multiply the following numbers by **(a)** 0.1 **(b)** 0.01 **(c)** 0.001:

16. 2700	**17.** 435	**18.** 1.7	**19.** 83.6	**20.** 4

Exercise 4.10 *Short division*

Do these divisions:

1. $4.6 \div 0.2$	**2.** $18.4 \div 0.4$	**3.** $2.85 \div 0.03$	**4.** $0.435 \div 0.05$
5. $2.148 \div 0.06$	**6.** $3.059 \div 0.7$	**7.** $0.1572 \div 0.03$	**8.** $0.333 \div 0.9$
9. $14.14 \div 0.7$	**10.** $9.12 \div 0.08$	**11.** $0.45 \div 0.2$	**12.** $6.34 \div 0.04$
13. $1.74 \div 0.8$	**14.** $0.0476 \div 0.7$	**15.** $37.4 \div 0.4$	**16.** $0.0711 \div 0.9$
17. $73 \div 0.5$	**18.** $1.64 \div 0.08$	**19.** $51 \div 0.03$	**20.** $0.008\,64 \div 0.6$

Exercise 4.11 *Conversion from fractions*

Write these as decimals:

1. $\frac{3}{10}$	**2.** $\frac{4}{5}$	**3.** $\frac{1}{2}$	**4.** $\frac{17}{100}$	**5.** $\frac{9}{20}$
6. $\frac{1}{4}$	**7.** $\frac{4}{25}$	**8.** $\frac{1}{5}$	**9.** $\frac{11}{50}$	**10.** $\frac{3}{20}$
11. $\frac{3}{50}$	**12.** $\frac{17}{25}$	**13.** $\frac{1}{10}$	**14.** $\frac{1}{100}$	**15.** $\frac{3}{4}$
16. $\frac{21}{25}$	**17.** $\frac{1}{20}$	**18.** $\frac{1}{8}$	**19.** $\frac{1}{3}$	**20.** $\frac{5}{9}$
21. $\frac{3}{8}$	**22.** $\frac{9}{11}$	**23.** $\frac{7}{16}$	**24.** $\frac{5}{12}$	**25.** $\frac{7}{8}$
26. $\frac{5}{6}$	**27.** $\frac{2}{3}$	**28.** $\frac{8}{11}$	**29.** $\frac{5}{8}$	**30.** $\frac{1}{12}$

Exercise 4.12 *Ordering*

Arrange in ascending order (start with the smallest):

1. $\frac{5}{8}, \frac{1}{2}, \frac{3}{4}$	**2.** $\frac{5}{6}, \frac{3}{4}, \frac{2}{3}$	**3.** $\frac{1}{3}, \frac{2}{5}, \frac{11}{30}$	**4.** $\frac{7}{12}, \frac{1}{2}, \frac{2}{3}$
5. $\frac{2}{3}, \frac{9}{16}, \frac{3}{4}$	**6.** $\frac{3}{5}$, 0.4, $\frac{7}{10}$	**7.** 0.75, $\frac{17}{20}$, 0.8	**8.** $\frac{4}{9}$, 0.4, $\frac{21}{50}$
9. $\frac{7}{10}$, 0.63, $\frac{17}{25}$	**10.** $\frac{4}{7}, \frac{4}{9}$, 0.5		

Arrange in descending order (start with the largest):

11. $\frac{7}{12}, \frac{1}{2}, \frac{2}{3}$	**12.** $\frac{9}{16}, \frac{3}{4}, \frac{5}{8}$	**13.** $\frac{3}{8}, \frac{1}{2}, \frac{3}{5}$	**14.** $\frac{1}{3}, \frac{1}{2}, \frac{5}{12}$
15. $\frac{3}{4}, \frac{1}{2}, \frac{3}{18}$	**16.** $\frac{1}{4}$, 0.2, $\frac{3}{10}$	**17.** $\frac{18}{25}$, 0.7, $\frac{3}{4}$	**18.** $\frac{1}{8}, \frac{3}{25}$, 0.1
19. $\frac{2}{3}, \frac{13}{20}$, 0.6	**20.** $\frac{2}{5}$, 0.45, $\frac{4}{9}$		

5

Measurement

Exercise 5.1 *Money*

1.	Write in pence:	(**a**)	£4	(**b**)	£20	(**c**)	£3.50	(**d**)	£2.05	(**e**)	£1.37
2.	Write in £:	(**a**)	600p	(**b**)	124p	(**c**)	642p	(**d**)	1550p	(**e**)	94p
		(**f**)	87p	(**g**)	8p	(**h**)	80p	(**i**)	6p	(**j**)	1600p

Add the following:

3. £1.63, 49p, £3.06
4. 72p, £1.83, 18p
5. 96p, 43p, 28p
6. 5p, £1.22p, 89p

Subtract the second amount of money from the first:

7. £1.50, £1.27
8. £2, £1.42
9. 82p, 17p
10. 62p, 47p
11. £8.32, £1.68
12. £1, 28p

Multiply:

13. 63p by 5
14. 7p by 8
15. 16p by 7
16. 22p by 6
17. £1.36 by 7
18. £3.16 by 4

Divide:

19. £0.81 by 3
20. 30p by 5
21. 36p by 4
22. £7.62 by 6
23. £6.80 by 8
24. £2.52 by 9
25. £7.26 by 3
26. £43.10 by 5
27. £3.42 by 9
28. £9.64 by 4
29. £52.08 by 6
30. £2.96 by 8

Exercise 5.2 *Money problems*

1. John has the following coins in his money-box:
3×50p, 12×10p, 7×5p, 19×2p and 8×1p.
How much money does he have altogether?

2. Steve has £4.50 and Sally has £2. How much must Steve give to Sally if they are to have the same amount of money each?

3. Adam receives 75p pocket money per week, Bill 50p and Caroline 25p.
(**a**) How much will they receive altogether in 12 weeks?
(**b**) How long will they take to receive £78?

4. Simon Jennings buys 4 pencils at 7p each and 2 exercise books at 17p each. How much does he spend altogether and how much change does he have from £1?

5. Mrs Jennings spent £3.00 at the greengrocer. She bought 6 kg of potatoes at 22p per kg, $1\frac{1}{2}$ kg of peas at 64p per kg, and a melon. What did the melon cost?
6. Carpet is sold at £5.25 per m^2. How much will it cost to buy (**a**) 6 m^2? (**b**) 17.4 m^2?
7. Six copies of a book cost £4.50. What does one book cost? What do 20 copies cost?
8. A dozen bottles of lemonade cost £3.84. What is the cost of a bottle?
9. A man earns £170 for working 40 hours. What does he earn in an hour?
10. In 1983 petrol cost £1.79 per gallon. A car travelled at 32 miles per gallon. How much did it cost to travel 960 miles?
11. A woman wants some new curtains which are priced at £30 in a shop. She decides to make her own and buys 4.5 m of material at £4.60 per metre. She spends another £4.20 on tape, cotton etc. How much does she save by making her own curtains?
12. Oranges are sold at 9p each or 6 for 52p. How much is saved if 60 oranges are bought at the cheaper rate?
13. In 1980 the postal rates were 12p for 1st class letters and 10p for 2nd class. How many more letters could be sent by 2nd class mail for £15 than could be sent by 1st class mail?
14. A taxi company charges 50p for the first km and 15p for each further 0.5 km. How much does it cost for a journey of 17.5 km?
15. The cost of making a telephone call to Blandford on a weekday in 1979 was 3p for:

    10 seconds between 9 a.m. and 1 p.m.
    15 seconds between 1 p.m. and 6 p.m.
    1 minute after 6 p.m.

    (**a**) What was the cost of a 4-minute call at noon?
    (**b**) What was the cost of a $6\frac{1}{2}$-minute call at 4 p.m.?
    (**c**) James made a 5 minute call at 8 p.m.
    What did it cost?
    For how long could he have made a call at 9.30 a.m. for the same price?

Exercise 5.3 *The metric system*

Convert these amounts to the given unit:

1.	To mm:	(**a**)	3 cm	(**b**)	5.3 cm	(**c**)	10.25 cm	(**d**)	0.6 cm
2.	To cm:	(**a**)	8 m	(**b**)	6.4 m	(**c**)	12 m	(**d**)	0.5 m
3.	To m:	(**a**)	7 km	(**b**)	4.8 km	(**c**)	20 km	(**d**)	0.75 km
4.	To mg:	(**a**)	7 g	(**b**)	45 g	(**c**)	0.5 g	(**d**)	13.5 g
5.	To g:	(**a**)	6 kg	(**b**)	2.2 kg	(**c**)	0.5 kg	(**d**)	0.05 kg
6.	To kg:	(**a**)	6 tonnes	(**b**)	14.5 tonnes	(**c**)	0.06 tonne	(**d**)	0.1 tonne
7.	To ml:	(**a**)	8 litres	(**b**)	4.5 litres	(**c**)	0.72 litre	(**d**)	0.005 litre
8.	To cm:	(**a**)	50 mm	(**b**)	73 mm	(**c**)	14.5 mm	(**d**)	7 mm
9.	To m:	(**a**)	600 cm	(**b**)	1450 cm	(**c**)	30 cm	(**d**)	80 cm
10.	To km:	(**a**)	2000 m	(**b**)	450 m	(**c**)	5 m	(**d**)	70 m
11.	To g:	(**a**)	7200 mg	(**b**)	500 mg	(**c**)	25 mg	(**d**)	5 mg
12.	To kg:	(**a**)	12 500 g	(**b**)	450 g	(**c**)	200 g	(**d**)	40 g
13.	To tonnes:	(**a**)	3000 kg	(**b**)	700 kg	(**c**)	250 kg	(**d**)	1 kg
14.	To litres:	(**a**)	6000 ml	(**b**)	47 ml	(**c**)	8 ml	(**d**)	750 cm^3

Do these conversions:

15. 10.7 cm to mm
16. 800 mm to cm
17. 1.5 m to cm
18. 700 cm to m
19. 8.4 km to m
20. 1500 m to km
21. 6000 mg to g
22. 4.3 g to mg
23. 600 g to kg
24. 8.75 kg to g
25. 4.5 tonnes to kg
26. 600 kg to tonnes
27. 1.375 l to ml
28. 115 cm to m
29. 2.54 cm to mm
30. 5500 m to km
31. 1.9 km to m
32. 0.7 kg to g
33. 6.25 l to cm^3
34. 500 mg to g
35. 1500 kg to tonnes
36. 7.5 m to cm
37. 8 mm to cm
38. 1.2 tonnes to kg

Exercise 5.4 *Fractions of quantities*

Calculate:

1. $\frac{1}{4}$ of £1
2. $\frac{3}{4}$ of £6
3. $\frac{2}{5}$ of 75 p
4. $\frac{3}{8}$ of £10
5. $\frac{4}{5}$ of £4.50
6. $\frac{3}{7}$ of £10.50
7. $\frac{5}{6}$ of £1.08
8. $\frac{5}{12}$ of £1.80
9. $\frac{1}{8}$ of £1.04
10. $\frac{2}{5}$ of 45p
11. $\frac{3}{4}$ of 1 m
12. $\frac{2}{3}$ of 4.5 m
13. $\frac{9}{10}$ of 6 cm
14. $\frac{3}{8}$ of 1 kg
15. $\frac{2}{5}$ of 1 tonne
16. $\frac{5}{8}$ of 1 litre
17. $\frac{5}{6}$ of 3 km
18. $\frac{4}{5}$ of 1 minute
19. $\frac{2}{3}$ of 1 revolution
20. $\frac{1}{6}$ of a right angle

Exercise 5.5 *Proportional parts*

Express the first quantity as a fraction of the second quantity. Give your answer in lowest terms:

1. 40p, £1
2. 6p, £2
3. 36p, £3
4. 45p, £6
5. 18p, £3.60
6. £1.60, £6.40
7. 52p, £10.40
8. 15p, 50p
9. 35p, £1.05
10. 8p, £2
11. 8 mm, 1 cm
12. 45 cm, 1 m
13. 9 mm, 1.5 cm
14. 85 cm, 4 m
15. 750 m, 2 km
16. 1.2 km, 6 km
17. 80 ml, 1 litre
18. 400 kg, 10 tonnes
19. 40 seconds, 1 minute
20. 25 min, 2 hours

Exercise 5.6 *Four operations involving metric units*

Add together the following quantities. Give your answer in the unit in brackets:

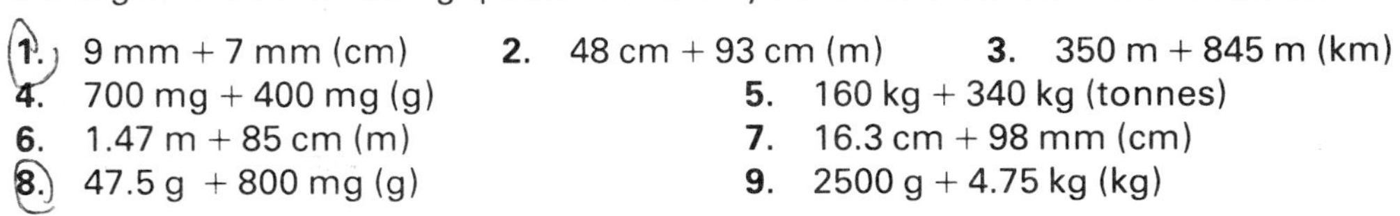

1. 9 mm + 7 mm (cm)
2. 48 cm + 93 cm (m)
3. 350 m + 845 m (km)
4. 700 mg + 400 mg (g)
5. 160 kg + 340 kg (tonnes)
6. 1.47 m + 85 cm (m)
7. 16.3 cm + 98 mm (cm)
8. 47.5 g + 800 mg (g)
9. 2500 g + 4.75 kg (kg)

Do the following subtractions. Give your answer in the unit in brackets:

10. 13 cm − 48 mm (cm)
11. 4.5 m − 435 cm (m)
12. 23.5 cm − 196 mm (cm)
13. 6 g − 140 mg (g)
14. 7.5 g − 1800 mg (g)
15. 8 kg − 1500 g (kg)

16. 6 tonnes − 750 kg (tonnes)
17. 900 kg − 0.25 tonnes (kg)
18. 4 litres − 250 ml (litres)

Multiply these amounts by the given number. Give your answer in the unit in brackets:

19. 9 mm × 11 (cm)
20. 3.7 mm × 7 (cm)
21. 85 cm × 35 (m)
22. 38.6 cm × 6 (m)
23. 350 m × 4 (km)
24. 450 mg × 3 (g)
25. 625 g × 3 (kg)
26. 120 kg × 25 (tonnes)
27. 280 ml × 9 (litres)

Divide these amounts by the given number. Give your answer in the unit in brackets:

28. 19.6 cm by 7 (cm)
29. 4 cm by 5 (mm)
30. 37.2 m by 6 (m)
31. 3 m by 4 (cm)
32. 1.45 km by 5 (m)
33. 16.5 km by 2 (km)
34. 6 kg by 5 (kg)
35. 7.5 tonnes by 15 (kg)
36. 1 litre by 20 (ml)
37. 5.4 litres by 6 (cm^3)
38. 3 kg by 12 (g)
39. 18 m by 72 (cm)

Exercise 5.7 *Problems*

1. A relay race consists of 1 leg of 1500 m, 2 legs of 800 m, 4 legs of 400 m, and 8 legs of 200 m. What is the total length of the race in km?
2. It takes 160 cm of string to tie up a box.
   (a) How many of these boxes can be tied up using a ball of string 50 m in length?
   (b) How much string is left over?
3. A cake has mass 2.5 kg. If it is shared equally among 20 children, what is the mass of the slice each child receives?
4. A 10p coin has a diameter of 28 mm.
   (a) How many of these coins would be needed to make a straight line 21 m long?
   (b) What is the total value of these coins?
5. A medicine spoon holds 5 cm^3. Tina is told to take 2 spoonfuls of cough mixture three times a day.
   (a) Will a half-litre bottle last her a fortnight?
   (b) If your answer to (a) is 'yes', how much is left over? If your answer is 'no', how much more is needed?
6. A man's walking pace is 90 cm long and he takes 110 paces each minute. Assuming he walks at this speed all the time, how many km has he walked in half an hour?
7. A packet of 500 sheets of paper is 6 cm thick. What is the thickness of a single sheet in mm?
8. A kitchen shelf is 2 m long and 20 cm wide. How many tins of beans with a diameter of 7.5 cm can be fitted on the shelf, if the base of each tin is completely on the shelf?
9. When a box is half empty it has mass 700 g; when full it has mass 1.2 kg. What is the mass of the box alone?
10. On a tin of weedkiller it says: 'To cover 5 m^2, dissolve 135 g in 5 l of water'.
    (a) How many 500 g tins are needed to cover an area of 65 m^2?
    (b) How many m^2 can be covered with one 500 g tin? Give your answer to the

nearest m^2.

(c) How much weedkiller should be dissolved in 3 l of water?

Exercise 5.8 *Time*

Write as times on the a.m./p.m. clock:

1. 0800
2. 1200
3. 1600
4. 2100
5. 0410
6. 0905
7. 2045
8. 1710
9. 1506
10. 2355

Write as times on the 24 hour clock

11. 7 a.m.
12. 1 p.m.
13. 10 p.m.
14. 8.15 a.m.
15. 11.06 a.m.
16. 12.30 p.m.
17. 2.25 p.m.
18. 8.05 p.m.
19. Quarter past three in the morning
20. Half past ten in the morning
21. Quarter to four in the afternoon
22. Twenty past seven in the morning
23. Ten to eleven in the morning
24. Twenty five past seven in the evening
25. Eighteen minutes to nine in the morning
26. Quarter to midnight
27. Five past four in the afternoon
28. Sixteen minutes to nine in the morning
29. Half past midnight
30. Quarter to twelve (noon)

Exercise 5.9 *Problems on time*

The following is an extract from a railway timetable:

Charing Cross	0810	
Sevenoaks	0848	Trains leave Charing Cross every hour.
Ashford	0922	
Canterbury	0948	Return trains from Margate leave at
Ramsgate	1007	20 minutes past each hours and take the
Margate	1018	same time as the down trains.

1. How long does the journey between Charing Cross and Margate take?
2. How long does it take to travel between Sevenoaks and Ramsgate?
3. At what time does the 1210 train from Charing Cross arrive at Canterbury?
4. Write down the times the 1420 train from Margate stops at all the stations to Charing Cross.
5. What train from Margate would you catch to arrive in London by 2100?

The following is an extract from a bus timetable to show the journey between the Bus Station and the Harbour:

Bus Station	0700	
Manchester Street	0710	
Clock Tower	0714	Buses leave the Bus Station every
Bootle Park	0720	20 minutes and take the same time
Town Hall	0726	to do the journey.
Railway Station	0730	
The Pig's Head	0741	
Harbour	0750	

6. How long does it take to travel from:
   (**a**) Manchester Street to the Pig's Head?
   (**b**) the Clock Tower to the Railway Station?
   (**c**) the Town Hall to the Harbour?
7. At what time will the 0820 bus arrive at:
   (**a**) Bootle Park?
   (**b**) the Pig's Head?
8. Mr Jones has an appointment at the Town Hall at 1000.
   Which bus must he catch from the Bus Station to arrive in time?
9. Mrs Smith wants to catch the 1115 train to London.
   What is the last bus she can catch from Manchester Street to arrive at the station in time?
10. It takes Jimmy Brown 8 minutes to walk to Bootle Park.
    If he leaves home at midday, how long will he have to wait before a bus arrives?
11. The White family are due to catch the 2000 boat from the Harbour.
    If they have to be at the Harbour an hour before, which bus must they catch from the Railway Station?

12. The 2300 ferry from Dover takes four and a quarter hours to reach Ostend. At what time does it arrive?
13. Mr Williams leaves home at 0740 and arrives at the office an hour and three quarters later. When does he arrive?
14. The film 'Schooldays are Fun' lasts for one hour and fifty five minutes. If it finishes at 2205, when does it start?
15. A school day consists of two 40-minute lessons and five 35-minute lessons. How long do a whole day's lessons last?

6

Factors

Exercise 6.1 *Prime numbers*

1. Which of the following numbers are prime?
   2, 3, 4, 7, 9, 21, 23, 27
2. Write down all the prime numbers less than 20.
3. Write out the numbers 1 to 50 arranged as shown below:

   1 2 3 4 5 6 7 8 9 10
   11 12 13 14 15 16 17 18 19 20
   21 etc.

   Now cross out all the multiples of 2, then all the multiples of 3 not already crossed out, and so on until you have crossed out all the numbers that are **not** prime.
4. Repeat question 3 for the numbers 51 to 100.

Exercise 6.2 *Products of factors*

Do these multiplications:

1. $2 \times 2 \times 3$	**2.** $2 \times 2 \times 2 \times 2$	**3.** $2 \times 3 \times 5$	**4.** $2 \times 2 \times 3 \times 3$
5. $2 \times 2 \times 3 \times 5$	**6.** $2 \times 3 \times 7$	**7.** $2 \times 5 \times 7$	**8.** $2 \times 3 \times 5 \times 7$
9. $2 \times 2 \times 5 \times 7$	**10.** $2 \times 5 \times 11$		

Calculate:

11. 2^3	**12.** 3^3	**13.** $2^2 \times 5$	**14.** 2^5
15. $2^3 \times 5$	**16.** 2^6	**17.** 2×3^2	**18.** $2 \times 3 \times 5^2$
19. $2^3 \times 5 \times 7$	**20.** $2^2 \times 3^2 \times 5 \times 7$		

Exercise 6.3 *Prime factors*

Write these as products of prime factors:

1. 8	**2.** 18	**3.** 20	**4.** 28	**5.** 30
6. 32	**7.** 42	**8.** 45	**9.** 48	**10.** 50
11. 54	**12.** 72	**13.** 80	**14.** 96	**15.** 108
16. 140	**17.** 144	**18.** 160	**19.** 450	**20.** 750

Exercise 6.4 *Prime factors in index form*

Write these as products of prime factors using index form:

1. 12	**2.** 16	**3.** 24	**4.** 27	**5.** 36
6. 40	**7.** 44	**8.** 52	**9.** 60	**10.** 64

11. 75 12. 84 13. 88 14. 98 15. 100
16. 120 17. 132 18. 180 19. 300 20. 850

Exercise 6.5 *Square roots*

Find the square root of each of these:

1. $2 \times 2 \times 2 \times 2$ 2. $2 \times 2 \times 17 \times 17$ 3. $2 \times 2 \times 3 \times 3 \times 11 \times 11$
4. $5 \times 5 \times 5 \times 5 \times 5 \times 5$ 5. $3 \times 3 \times 3 \times 3 \times 7 \times 7$
6. 36 7. 64 8. 81 9. 100 10. 144

Express the following numbers as products of prime factors and hence find their square roots:

11. 196 12. 400 13. 256 14. 225
15. 441 16. 576 17. 324 18. 784
19. 625 20. 729 21. 484 22. 900
23. 1225 24. 1600 25. 676 26. 1024
27. 1936 28. 1296 29. 2304 30. 1764

What is the least number by which the following must be multiplied to make them perfect squares?

31. 48 32. 80 33. 160 34. 216 35. 396

What is the least number by which the following must be divided to make them perfect squares?

36. 72 37. 147 38. 405 39. 1350 40. 2560

Exercise 6.6 *Highest Common Factor*

Find the Highest Common Factor of each of the following groups of numbers:

1. 2×3 and 2×3^2 2. $2^2 \times 3^2$ and $2^3 \times 3$ 3. $2 \times 3 \times 5$ and $2^2 \times 5$
4. $2 \times 3 \times 5 \times 7$ and $2 \times 3^2 \times 7$ 5. $2^3 \times 3^2 \times 5$ and $2^3 \times 3 \times 5^2$
6. $2^3 \times 3 \times 7$ and $2 \times 3^3 \times 5$ and $2 \times 3^2 \times 7$
7. $3^3 \times 5 \times 11$ and 3×5^2 and $2 \times 5 \times 11$
8. $2^3 \times 3 \times 7$ and $2 \times 3 \times 5$ and $2^4 \times 13$
9. $2^3 \times 7$ and $3^2 \times 5$ and 5^3
10. 12 and 18 11. 36 and 48 12. 18 and 30
13. 50 and 120 14. 42 and 70 15. 120 and 180
16. 12 and 30 and 36 17. 28 and 42 and 70
18. 48 and 120 and 168 19. 70 and 105 and 175
20. 9 and 16 and 25

Exercise 6.7 *Lowest Common Multiple*

Find the Lowest Common Multiple of each of the following groups of numbers:

1. 2×3 and 2×3^2 2. $2^2 \times 3^2$ and $2^4 \times 3$
3. $2 \times 3 \times 5$ and $2^2 \times 5$ 4. $2^2 \times 3 \times 5$ and $2^2 \times 5^2$
5. $2^2 \times 3, 2^3 \times 3, 2 \times 3^2$ 6. $2^2 \times 3 \times 5, 2 \times 3 \times 5^2, 2^2 \times 3^2 \times 5$

7. $2 \times 3 \times 5, 2^3 \times 3, 3 \times 5 \times 7$

8. $2 \times 5 \times 7, 2^2 \times 7, 3 \times 5 \times 7$

9. 12 and 18

10. 24 and 36

11. 18 and 30

12. 50 and 60

13. 42 and 70

14. 63 and 84

15. 12, 18, 27

16. 12, 30, 36

17. 16, 20, 30

18. 18, 24, 27

19. 36, 48, 108

20. 90, 120, 180

Exercise 6.8 *Lowest Common Multiple – problems*

1. 3 bells start chiming at the same time. The first chimes every 6 seconds, the second every 8 seconds and the third every 10 seconds. After how long do they chime together again?

2. Two lights flash at intervals of 6 seconds and 15 seconds respectively. If they start flashing together, after how long will they next flash together?

3. What is the smallest number of stamps a dealer can divide into packets of either 15, 20 or 25?

4. Nicholas remembers that his telephone number is the first number that can be divided by his age, 12, his mother's age, 40, and his father's age, 44. What is his telephone number?

5. 3 guns in a battery fire at different rates: at intervals of 1 minute, 1 minute 20 seconds and $1\frac{1}{2}$ minutes. If they start firing together, when will they next fire together?

7

Sets and Venn diagrams

Introduction – set symbols

In this chapter, the following symbols are used:

{ } : a set which is a collection of elements or members
ℰ : the universal set, i.e. the set that contains all elements under consideration
∈ : is a member of the set
n(A) : the number of elements in set A
⊂ : is a subset of
⊃ : contains as a subset

∪ : union of sets

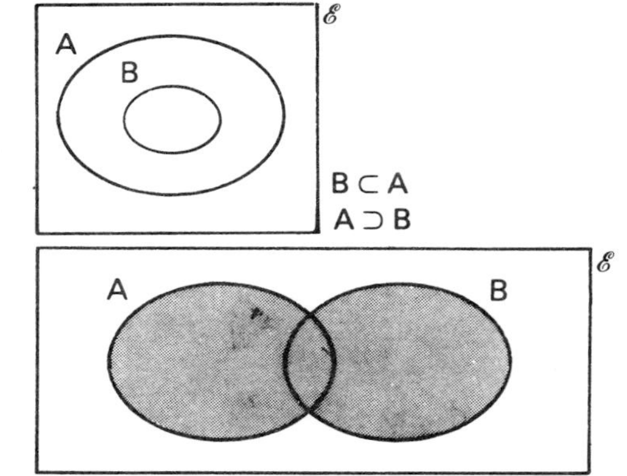

B ⊂ A
A ⊃ B

A ∪ B is the shaded region,
i.e. members of A or B or both.

∩ : intersection of sets

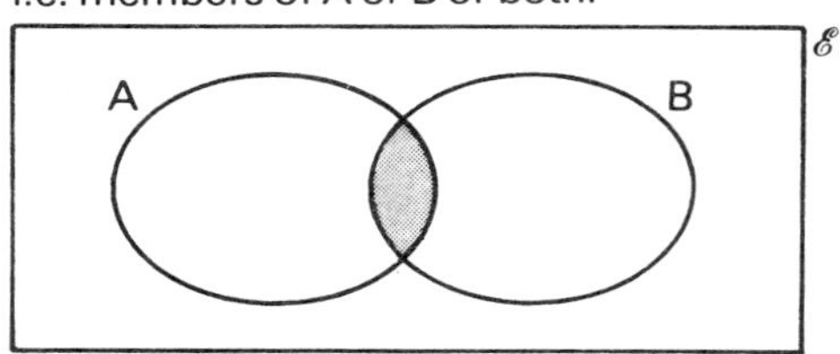

A ∩ B is the shaded region,
i.e. members of both A and B.

ø : the empty or null set

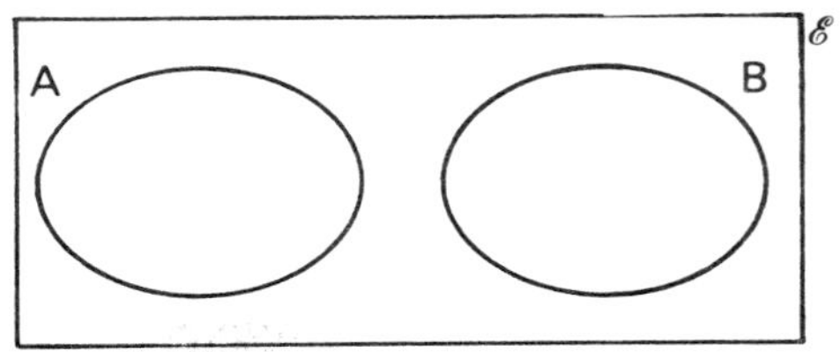

A ∩ B = ø
A and B are disjoint sets.

′ : the complement of a set

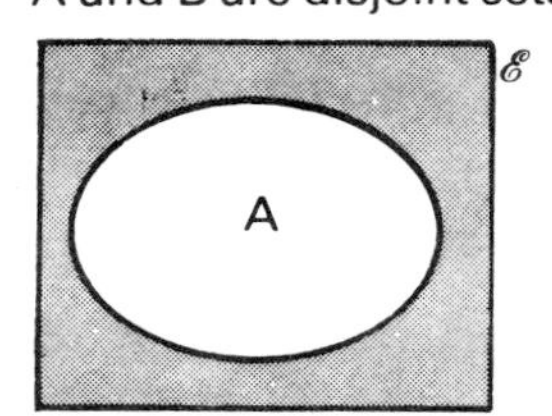

A′ is the shaded region,
i.e. all members of ℰ not in A.

Exercise 7.1 *Listing members*

List the members of the following sets:

1. A = {vowels in the alphabet}
2. B = {colours in the Union Jack}
3. C = {months of the year starting with the letter A}
4. D = {suits in a pack of playing cards}
5. E = {forms in your school}
6. F = {positive whole numbers less than 6}
7. G = {negative whole numbers greater than -4}
8. H = {prime numbers < 13}
9. I = {whole numbers that divide into 15 without leaving a remainder}
10. J = {the first 4 perfect squares}

Exercise 7.2 *The universal set*

Write down suitable names for the universal set $\mathscr{E}$:

1. {Alan, Brian, Charles, David}
2. {rose, daffodil, daisy}
3. {beer, lemonade, wine}
4. {tennis, netball, cricket}
5. {triangle, square, pentagon}
6. {square, kite, parallelogram}
7. {binary, octal, denary}
8. {1, 2, 3, 4}
9. {1, 2, 3, . . ., 10}
10. {1, 3, 5, 7, 9}
11. {2, 4, 6, 8}
12. {3, 6, 9, . . ., 21}
13. {5, 10, 15, 20, 25}
14. {1, 4, 9, 16}
15. $\{\frac{1}{2}, \frac{1}{3}, \frac{1}{4}\}$

Exercise 7.3 *Membership*

Use the symbol $\in$ or $\notin$ to show whether the element named is, or is not, a member of the given set. Use the letter given in brackets to represent the element.

Example: William (w) K = {Kings of England}
Answer: $w \in K$

1. George (g) N = {boys' names}
2. π (p) A = {letters of the English alphabet}
3. 4 (f) P = {positive integers}
4. Italy (i) C = {European countries}
5. Edinburgh (e) T = {towns in England}
6. Two (t) P = {prime numbers}
7. Rhombus (r) T = {triangles}
8. Cube (c) S = {solid figures}
9. 7 (s) P = {positive even integers}
10. 8 (e) M = {multiples of 2}

Exercise 7.4 *Number of elements*

What is the number of elements in the following sets?

Example: X = {finger holes on a telephone dial}
Answer: $n(X) = 10$

1. A = {days of a week}

2. B = {playing cards in a pack without jokers}
3. C = {letters in the alphabet}
4. D = {months with 30 days}
5. E = {tentacles of an octopus}
6. F = {positive multiples of 4 less than 16}
7. G = {positive integers < 8}
8. H = {positive integers $\leqslant 7$}
9. I = {prime numbers between 6 and 12}
10. J = {positive multiples of 6 less than 5}

In questions 11–15 r is a whole number

11. K = $\{6 < r < 10\}$
12. L = $\{-4 > r > -10\}$
13. M = $\{0 < r \leqslant 15\}$
14. N = $\{10 \leqslant r \leqslant 15\}$
15. O = $\{r > 0\}$

Exercise 7.5 *Subsets*

Draw Venn diagrams to show the relationship between each pair of sets. Write down the connection between the two using $\subset$ or $\supset$.

Example: A = {1, 2, 3, 4, 5} B = {1, 2, 3}
Answer: A $\supset$ B

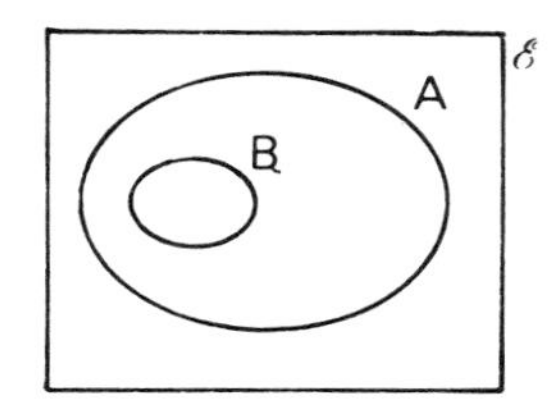

1. A = {Susan, Linda, David, Jackie} B = {Susan, Linda, Jackie}
2. C = {James, John} D = {Robert, James, Richard, John, Roger}
3. E = {a, b, c} F = {a, b, c, d, e, f}
4. G = {spaniel, beagle, poodle} H = {spaniel}
5. I = {vowels} J = {letters of the alphabet}
6. K = {positive integers < 10} L = {2, 4, 6, 8}

Exercise 7.6 *Relationships between two sets*

Draw Venn Diagrams to show the relationship between each pair of sets, putting each element in its correct region:

1. A = {1, 2, 3, 4, 5} B = {4, 5, 6, 7}
2. C = {2, 4, 6, 8} D = {3, 6, 9, 12}
3. E = {a, b, c, d, e} F = {b, c, d}
4. G = {k, l, m, n} H = {l, m, n, o}
5. I = {Liverpool, Arsenal, Everton} J = {Manchester City, Stoke, Brighton}
6. K = {R, S, V, P} L = {S, P, Q, R}
7. M = {a, e, i, o, u} N = {b, c, d, f}
8. O = {K, C, M, G} P = {C, M, G}
9. Q = {positive integers < 10} R = {first 2 positive multiples of 5}
10. S = {positive integers < 10} T = {first 5 positive multiples of 2}
11. U = {positive integers < 10} V = {negative integers > -5}
12. W = {positive integers < 10} X = {2, 3, 5, 7}

Exercise 7.7 *Venn diagrams – regions*

1. Copy the diagram four times. Shade in these regions, using a different copy for each:
   (**a**) $A \cap B$
   (**b**) $A \cup B$
   (**c**) $A \cap B'$
   (**d**) A

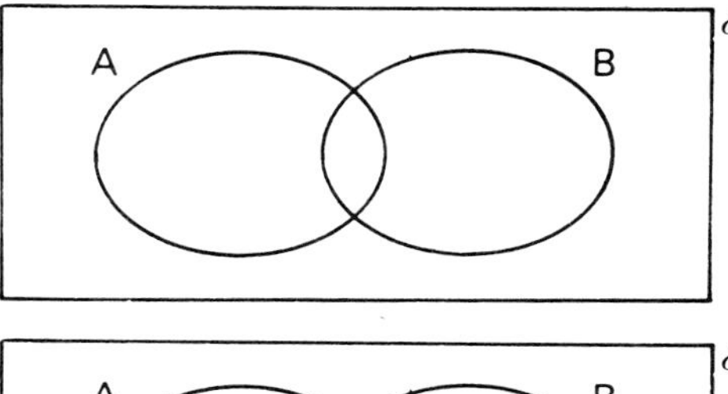

2. Copy the diagram four times. Shade in these regions, using a different copy for each:
   (**a**) $A' \cap B$
   (**b**) $(A \cup B)'$
   (**c**) $(A \cap B)'$
   (**d**) A'

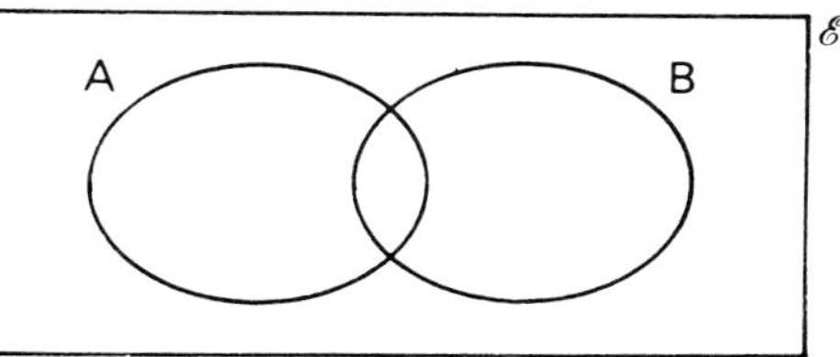

3. Name each of the shaded regions, using set language:

(**a**)

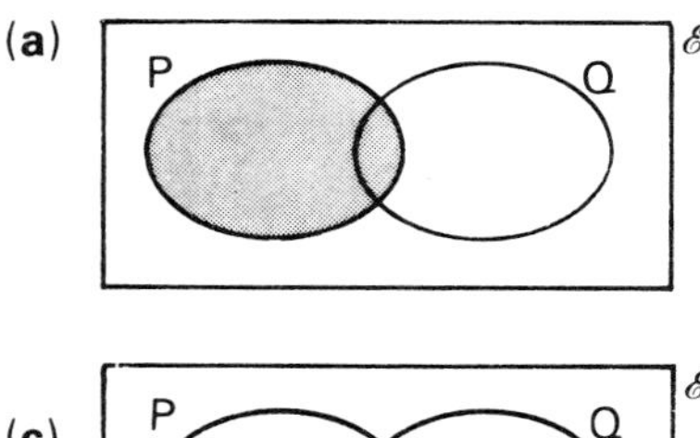

(**b**)

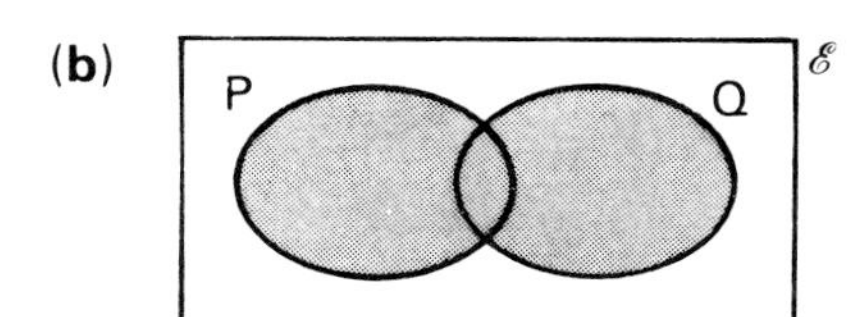

(**c**)

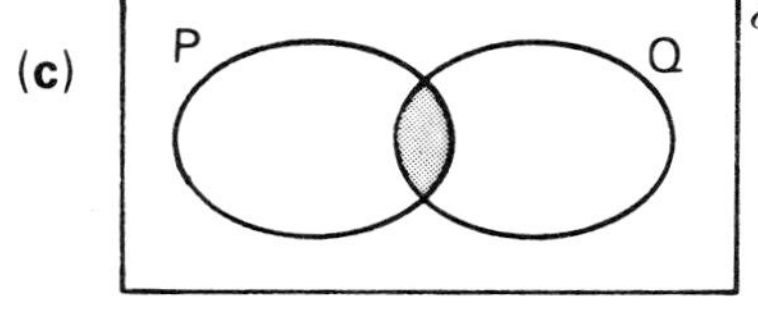

(**d**)

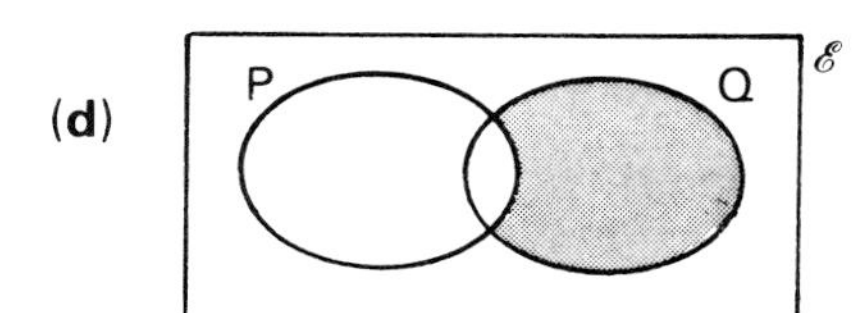

(**e**)

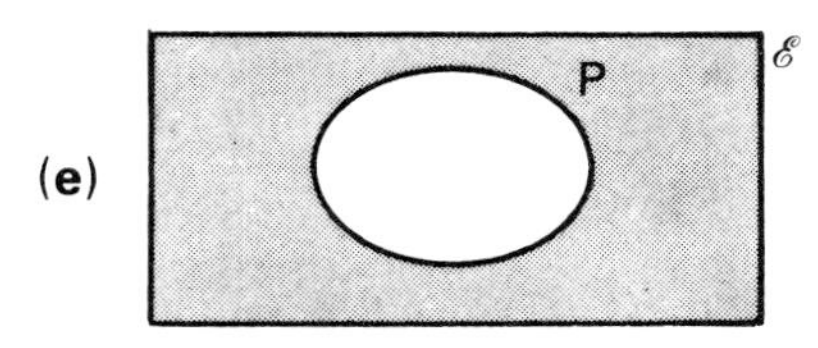

(**f**)

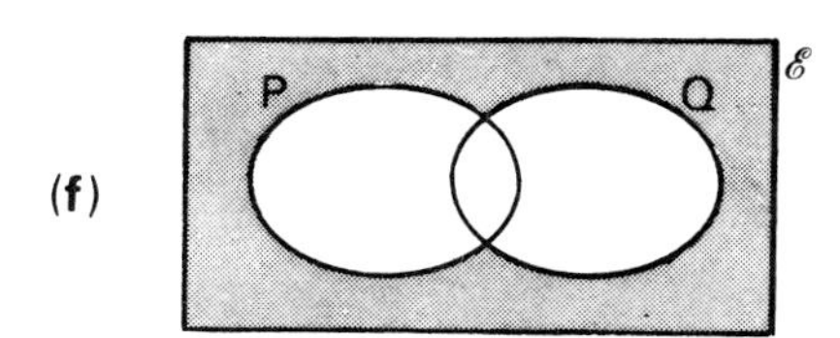

4. Copy the diagram three times. Shade in these regions, using a different copy for each:
   (**a**) $A \cap B$
   (**b**) $A \cup B$
   (**c**) $A' \cap B$

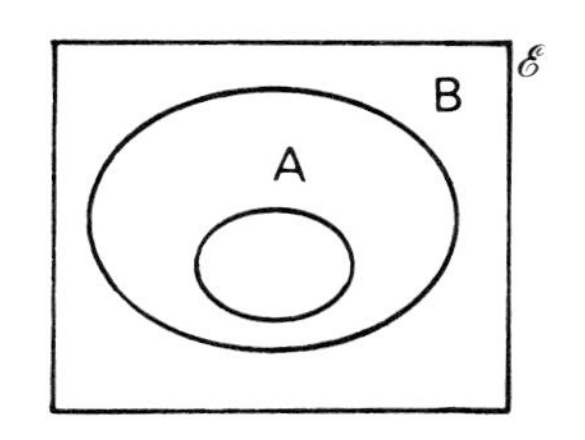

5. Copy the diagram three times. Shade in these regions, using a different copy for each:
   (**a**) B
   (**b**) $A \cup B$
   (**c**) $A \cap B'$

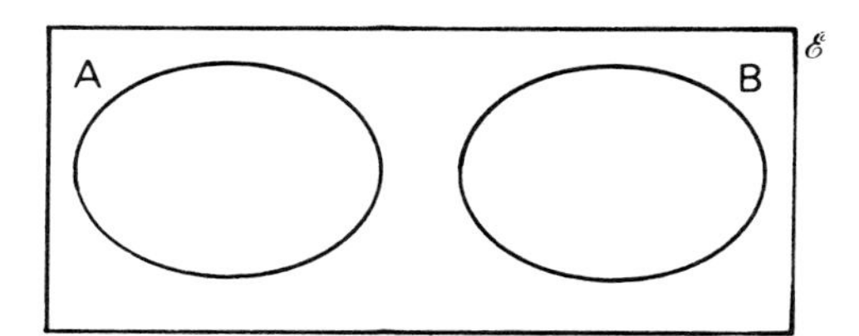

6. Draw diagrams to illustrate
   (a) $M \cap N$
   (b) $M \cup N$
   (c) $M' \cap N$

7. Copy the diagram three times. Shade in these regions, using a different copy for each:
   (a) $(X \cap Y)'$
   (b) $(X \cap Y) \cup (X \cap Y')$
   (c) $(X \cup Y) \cup (X \cup Y)'$

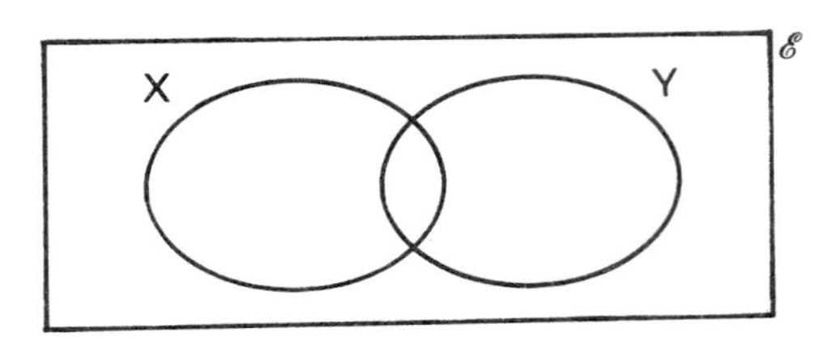

8. Name each of the shaded regions, using set language:

(a)

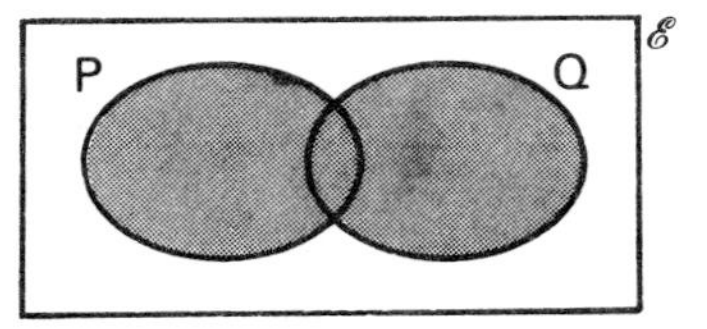

(b)

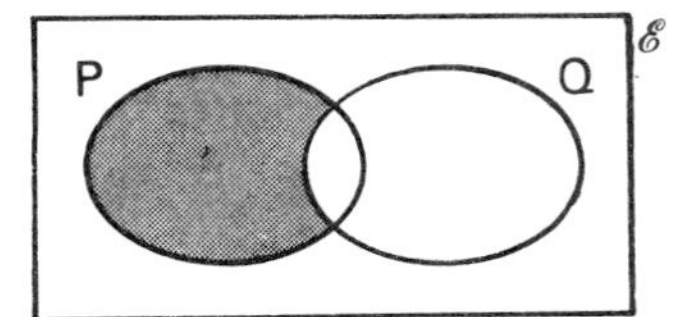

(c)

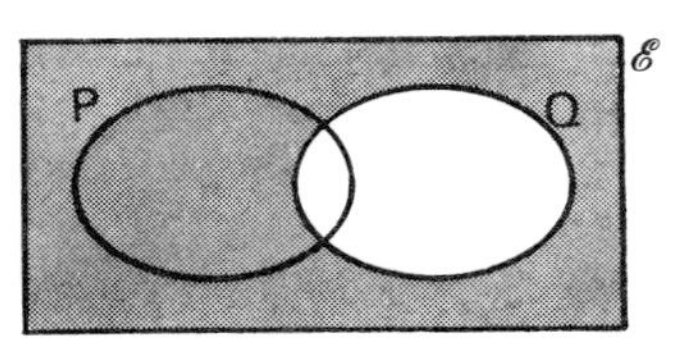

(d)

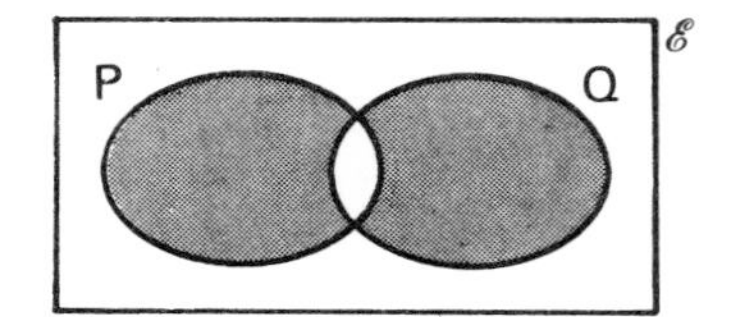

9. In the diagram,
   $\mathscr{E} = \{\text{positive integers} \leq 10\}$.
   (a) What is (i) $n(A)$ (ii) $n(A \cup B)$ (iii) $n(A \cap B')$?
   (b) What are the members of (i) $A \cap B$ (ii) $(A \cup B)'$ (iii) $A' \cap B$

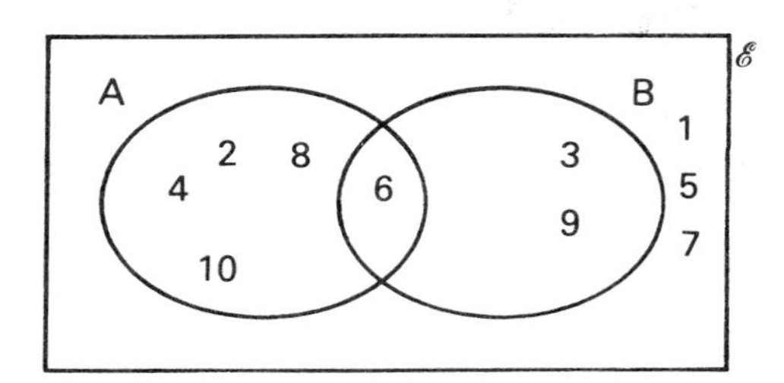

10. Copy the diagram.
    (a) Shade $P' \cap Q$ horizontally ≡
    (b) Shade $P \cap Q$ vertically ||||
    (c) Shade $(P \cup Q)'$ diagonally #

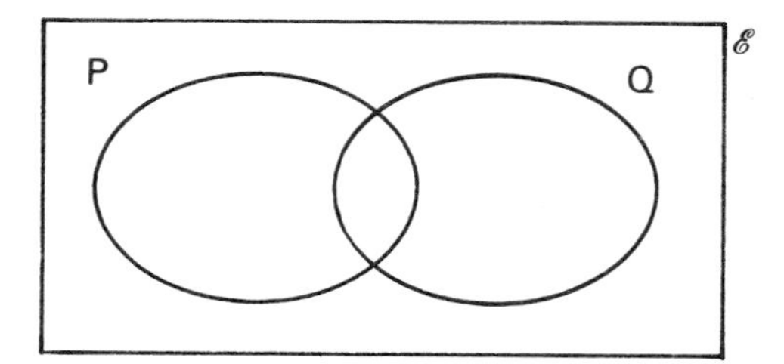

Exercise 7.8 *Problems involving Venn diagrams – two sets*

1. The figures show the **number** of elements in each region of the Venn diagram.
What is:
(**a**) $n(X)$ (**b**) $n(X \cap Y)$
(**c**) $n(X \cup Y)$ (**d**) $n(X \cap Y')$
(**e**) $n(X \cup Y)'$ (**f**) $n(\mathscr{E})$

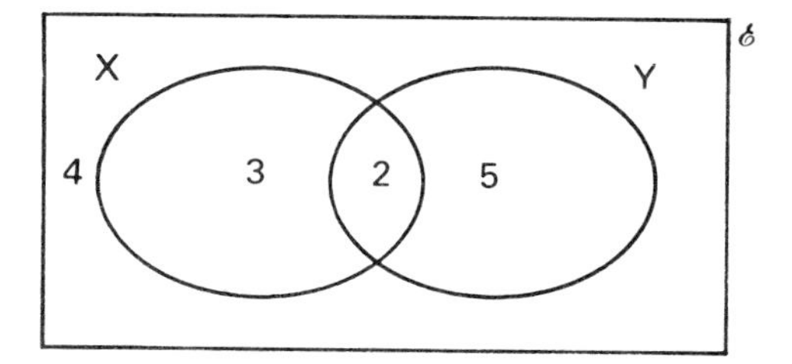

2. $\mathscr{E} = \{a, b, c, d, e, f\}$
$X = \{a, b, c\}$
$Y = \{a, e\}$
Copy the diagram and put each letter in its correct region.
How many elements are there in
(**a**) $X \cap Y$
(**b**) $X \cup Y$?

3. $\mathscr{E} = \{$Paul, Quentin, Robert, Sally, Tony$\}$
A = $\{$Paul, Quentin, Sally$\}$
B = $\{$Sally, Tony$\}$
Draw a Venn diagram to show the relationship between $\mathscr{E}$, A and B, putting the names in the correct region.

4. $\mathscr{E}$ = $\{$days of the week$\}$
L = $\{$those days that have 6 letters$\}$
S = $\{$those days that start with the letter S$\}$
(**a**) Draw a Venn diagram to show in which region each day belongs.
(**b**) How many days have 6 letters and start with S?
(**c**) How many days do not have 6 letters?
(**d**) What can you say about those days in $(L \cup S)'$?

5. $\mathscr{E} = \{1, 2, 3, 4, 5, 6, 7, 8, 9\}$
$A = \{1, 4, 9\}$
$B = \{2, 4, 6, 8\}$
(**a**) Draw a Venn diagram to show $\mathscr{E}$, A and B and put the numbers in their correct region.
(**b**) What is (**i**) $n(A \cap B)$ (**ii**) $n(A \cup B)$ (**iii**) $n(A \cup B)'$?
(**c**) Which of the following statements are true?
(**i**) $B \subset A$ (**ii**) $n(\mathscr{E}) = 45$ (**iii**) $1 \notin B$?

6. If $\mathscr{E} = \{a, b, c, d, e\}$
$A = \{a, b, c\}$
$B = \{b, d\}$
(**a**) What are the members of (**i**) $(A \cup B)$ (**ii**) A'?
(**b**) How many members are there in (**i**) $(A \cap B)$ (**ii**) $(A \cap B')$?
(**c**) What is the name given to the subset in which e is found?

7. $\mathscr{E} = \{A, B, C, D, E, F\}$
X = $\{$letters with a vertical axis of symmetry$\}$
Y = $\{$letters with a horizontal axis of symmetry$\}$
What are the members of (**a**) X (**b**) $X \cap Y$ (**c**) $(X \cup Y)'$

8.

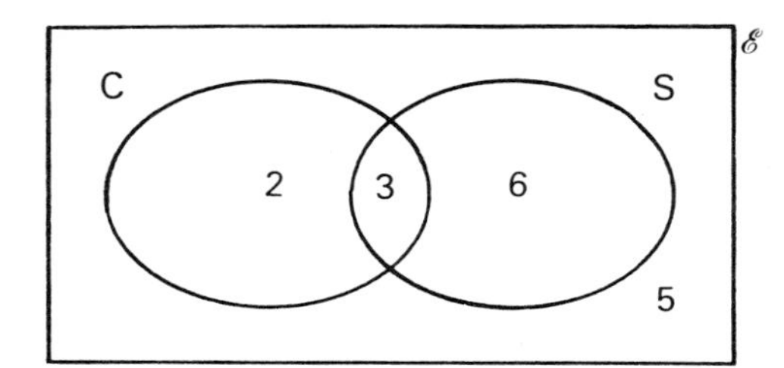

$\mathscr{E}$ = {members of form 3}
C = {those who collect cars}
S = {those who collect stamps}

(**a**) How many are there in form 3?
(**b**) How many collect cars?
(**c**) How many collect both cars and stamps?
(**d**) How many collect stamps but not cars?
(**e**) How many car collectors do not collect stamps?
(**f**) How many collect neither stamps nor cars?
(**g**) How many do not collect stamps?

9. (**a**) Show from the diagram the truth of the statement $n(A \cup B) = n(A) + n(B) - n(A \cap B)$

(**b**) Arrange the equation to make (**i**) $n(A \cap B)$ (**ii**) $n(A)$ the subject.

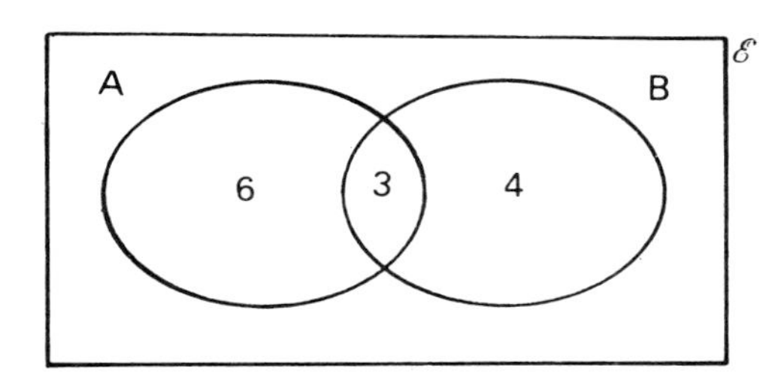

10. The boarders at a school have a choice of ice cream or jelly for Sunday lunch. The greedy ones can have both.
One Sunday when every pupil was present 85 ate ice cream, 43 chose jelly and 20 had both. How many boarders were there at the school?

11. All members of a family, except Granny, watched television one night. 4 watched a BBC programme but 3 of these watched an ITV programme as well, while 2 watched ITV only. How many are there in the family?

12. Of the hundred boys in a school, all either swim or play cricket while some do both. 80 play cricket and half the school swims. How many both swim and play cricket?

13. Draw sets A and B to illustrate the following information, putting the number of elements in each region.

(**a**) $n(A) = 6$
$n(A \cup B) = 10$
$n(A \cap B) = 4$

(**b**) $n(A) = 6$
$n(A \cup B) = 6$
$n(A \cap B) = 4$

(**c**) $n(A) = 6$
$n(B) = 4$
$n(A \cup B) = 10$

14. (**a**) Find $n(A \cap B)$ when $n(A) = 10$, $n(B) = 12$, $n(A \cup B) = 18$
(**b**) Find $n(A \cup B)$ when $n(A) = 10$, $n(B) = 12$, $n(A \cap B) = 6$

15. (**a**) Describe in set notation the relationship between A and B.
(**b**) If $n(A) = 10$ and $n(B) = 7$ find (**i**) $n(A \cup B)$
(**ii**) $n(A \cap B)$

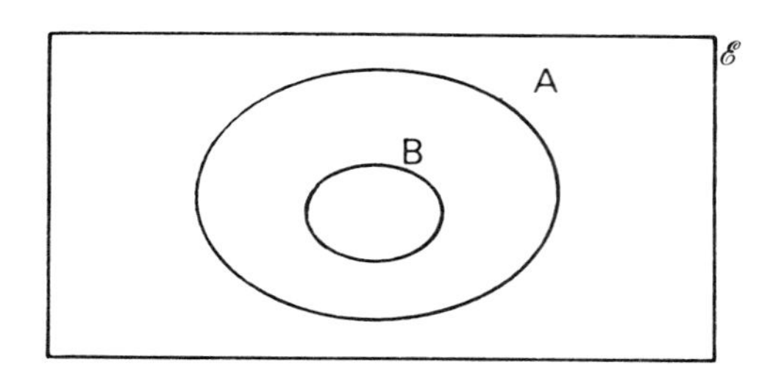

16. (**a**) What is (**i**) $n(P \cup Q)$
(**ii**) $n(P \cap Q)$?

(**b**) Copy and complete: $P \cap Q =$

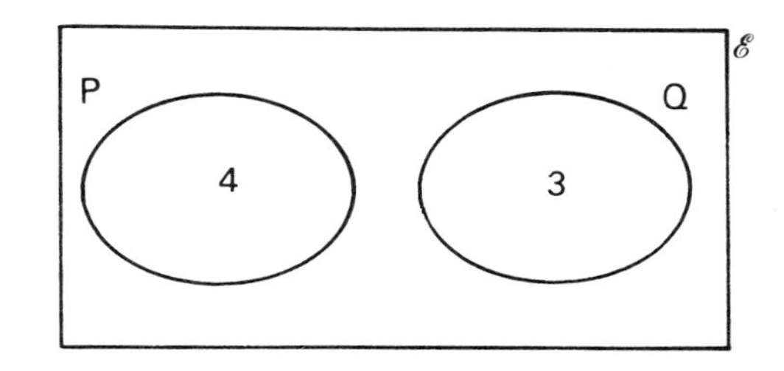

17. If $n(A) = 10$ and $n(B) = 6$:

(**a**) Draw Venn diagrams to show

(**i**) the greatest possible value of $n(A \cup B)$

(**ii**) the least possible value of $n(A \cup B)$.

(**b**) If $n(A \cap B) = 2$, how many elements are there in $(A \cup B)$?

18. Of 100 boys in a school, 75 have red ties and 50 have blue socks. What is

(**a**) the maximum number who can have both?

(**b**) the minimum number who can have both?

19. There are 10 librarians in a school and the school choir has 20 members.

(**a**) Draw Venn diagrams to show that the total number of pupils involved lies between 20 and 30.

(**b**) If there are 24 pupils involved, what conclusion can you draw?

20. 100 people were asked about their reading habits. 68 said they read novels, 52 said they liked biographies and 25 said they read both. What conclusion can you draw from these figures?

8

Perimeter and area

Exercise 8.1 *Squares*

1. Find the area of a square with the given side:

(a) 1 cm	(b) 5 cm	(c) 4 m	(d) 8 mm
(e) 3 km	(f) 12 cm	(g) 20 cm	(h) 24 km
(i) 2.5 m	(j) 0.5 m	(k) 0.3 km	(l) 0.25 m

2. Find the perimeters of the squares in Question 1.

Find the area of a square with the given side. Give your answer in the unit in brackets.

Worked examples: 5 cm (mm^2). 5 cm = 50 mm; $(50\text{ mm})^2 = 2500\text{ mm}^2$
60 cm (m^2). 60 cm = 0.6 m; $(0.6\text{ m})^2 = 0.36\text{ m}^2$

3. 4 cm (mm^2)	**4.** 2 cm (mm^2)	**5.** 3 km (m^2)	**6.** 0.8 cm (mm^2)
7. 0.4 m (cm^2)	**8.** 2.4 km (m^2)	**9.** 40 cm (m^2)	**10.** 15 mm (cm^2)
11. 120 cm (m^2)	**12.** 120 mm (cm^2)	**13.** 0.5 cm (m^2)	**14.** 18 m (km^2)

15. How many square millimetres are there in a square centimetre?
16. How many square centimetres are there in a square metre?
17. How many square metres are there in a square kilometre?

Find the area of a square with the given side. Then give the answer in each of the other units requested.

Worked example: 20 cm (cm^2, mm^2, m^2)
$20\text{ cm} \times 20\text{ cm} = 400\text{ cm}^2$. $400\text{ cm}^2 = 400 \times 100\text{ mm}^2 = 40\,000\text{ mm}^2$
$400\text{ cm}^2 = 400 \times 0.0001\text{ m}^2 = 0.04\text{ m}^2$

18. 30 cm (cm^2, mm^2, m^2)	**19.** 50 cm (cm^2, mm^2, m^2)
20. 80 cm (cm^2, mm^2, m^2)	**21.** 120 cm (cm^2, mm^2, m^2)
22. 20 m (m^2, cm^2, km^2)	**23.** 15 m (m^2, cm^2, km^2)
24. 8 mm (mm^2, cm^2)	**25.** 25 mm (mm^2, cm^2)
26. 0.4 m (m^2, cm^2, mm^2)	**27.** 0.35 m (m^2, cm^2, mm^2)
28. 45 cm (cm^2, mm^2, m^2)	**29.** 2.5 m (m^2, cm^2, dm^2)

Find (**a**) the length of a side, (**b**) the area, of a square with the given perimeter:

30. 12 cm	**31.** 20 m	**32.** 44 mm
33. 1.2 m	**34.** 6.4 cm	**35.** 2.4 km
36. 0.6 cm	**37.** 1600 m	**38.** 800 cm

Find the length of a side of a square with the given area:

39. 9 mm^2 **40.** 25 m^2 **41.** 4 km^2
42. 100 km^2 **43.** 144 m^2 **44.** 1.44 cm^2
45. 0.25 cm^2 **46.** 0.36 km^2 **47.** 784 cm^2

Exercise 8.2 *Rectangles*

1. Find the area of a rectangle with the given sides:
(**a**) 5 cm, 10 cm (**b**) 8 cm, 3 cm (**c**) 6 m, 2 m
(**d**) 7 m, 4 m (**e**) 10 m, 8 m (**f**) 15 cm, 4 cm
(**g**) 24 m, 6 m (**h**) 25 cm, 18 cm (**i**) 42 m, 15 m
(**j**) 1.2 km, 0.8 km (**k**) 6.5 cm, 4.8 cm (**l**) 0.36 m, 2.5 m

2. Find the perimeters of the rectangles in Question 1.

3. Find the area of a table 2 m long and 1.5 m wide.
Write the answer (**a**) in m^2 (**b**) in cm^2.

4. Find the area of a wall 8 m long and 2 m high.
Write the answer (**a**) in m^2 (**b**) in cm^2.

5. Find the area of a piece of paper 30 cm long and 21 cm wide.
Write the answer (**a**) in cm^2 (**b**) in m^2.

6. Find the area of a postage stamp 2.0 cm long and 2.5 cm wide.
Write the answer (**a**) in cm^2 (**b**) in mm^2.

7. A special Christmas issue stamp has length 4.4 cm and width 2.5 cm.
Find its area (**a**) in cm^2 (**b**) in mm^2.

8. Find the area of a rectangular cornfield 250 m long and 200 m wide
(**a**) in m^2
(**b**) in km^2.
(**c**) A hectare is $10\,000 \text{ m}^2$. What is the area of the cornfield in hectares?

9. A soccer pitch is 110 m long and 80 m wide. Find its area
(**a**) in m^2 (**b**) in km^2 (**c**) in hectares.

10. A rectangle has length 5 m. Its area is 30 m^2.
(**a**) Find its width.
(**b**) Hence calculate its perimeter.

Use the method of Question 10 to find the perimeter of each of the following rectangles:

11. length 4 m, area 12m^2 **12.** length 6 m, area 12 m^2
13. length 2 m, area 2 m^2 **14.** length 15 cm, area 75 cm^2
15. length 25 cm, area 175 cm^2 **16.** length 4 m, area 2 m^2
17. width 0.8 m, area 2 m^2 **18.** length 0.8 m, area 0.4 m^2
19. length 1.2 m, area 1.8 m^2 **20.** width 18 cm, area 576 cm^2

21. A rectangle has length 7 cm and perimeter 22 cm.
(**a**) Find its width.
(**b**) Hence calculate its area.

Use the method of Question 21 to find the area of each of the following rectangles:

22. length 6 cm, perimeter 18 cm
23. length 8 m, perimeter 26 m
24. length 5 km, perimeter 16 km
25. width 2 cm, perimeter 22 cm
26. width 16 cm, perimeter 112 cm
27. perimeter 80 cm, length 35 cm
28. perimeter 2 m, length 0.8 m
29. perimeter 1.8 m, length 0.5 m
30. perimeter 6 m, width 1.4 m
31. perimeter 8.5 m, width 0.25 m

32. An envelope has length 12 cm and area 102 cm^2.
Find (**a**) its width in cm (**b**) its perimeter in m.

33. A postcard has length 14 cm and perimeter 46 cm.
Find its area in cm^2.

34. A piece of sticking plaster has width 4.8 cm and area 180 cm^2.
Find (**a**) its length in cm (**b**) its perimeter in m.

35. A rugby field has length 90 m and area 5850 m^2.
Find (**a**) its width in m (**b**) its perimeter in km.

36. I run 12 times round the field in Question 35. How far do I run? Answer in km.

37. As a punishment, a boy is told to run at least 3 km. How many complete circuits of the field in Question 35 must he make?

Exercise 8.3 *Rectangular figures*

1. Find the perimeter of each of the following rectangular figures:

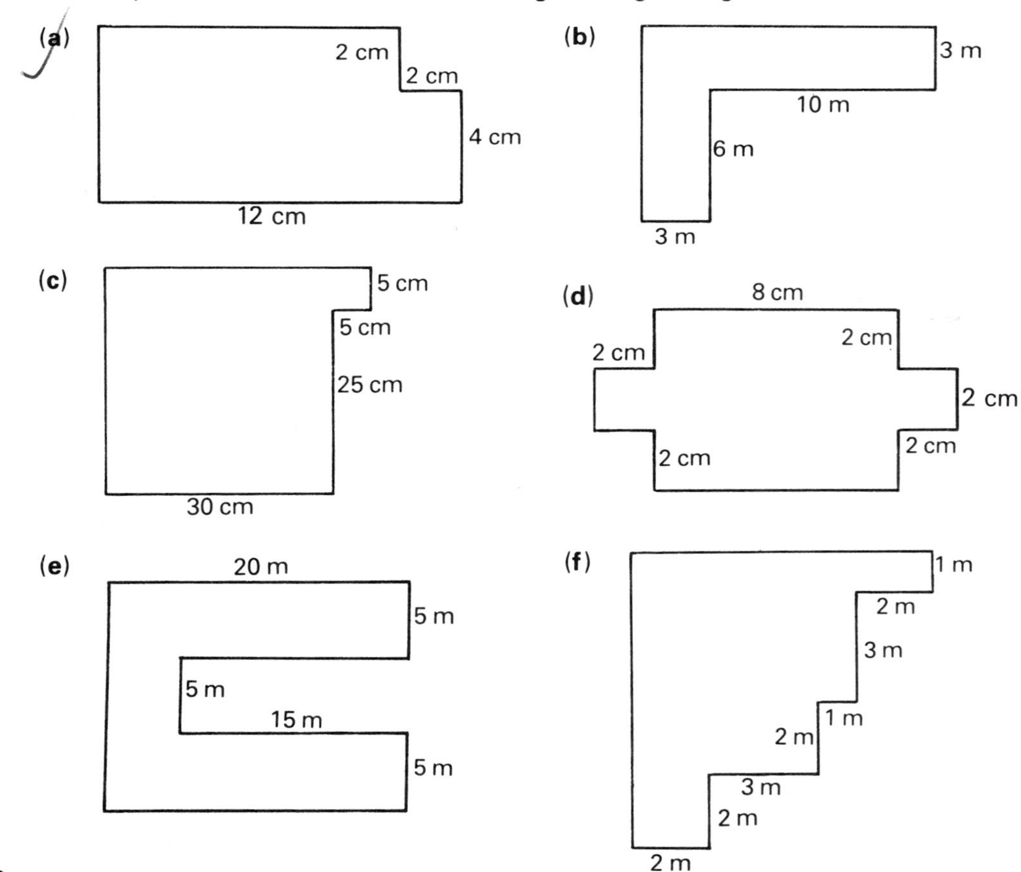

2. Find the areas of the figures in Question 1.

3. From each corner of a postcard a square with side 2 cm is removed.
   (a) How will this alter the perimeter of the card?
   (b) By how much will the area be reduced?
   (c) The card was originally 15 cm long and 10 cm wide. Find its final perimeter and area.

4. For a display board, letters are cut in rectangular shapes as illustrated below. Each letter is cut from a piece of cardboard originally 10 cm high and 6 cm wide. The figure is marked in 2 cm squares. Find the area of cardboard in each letter.

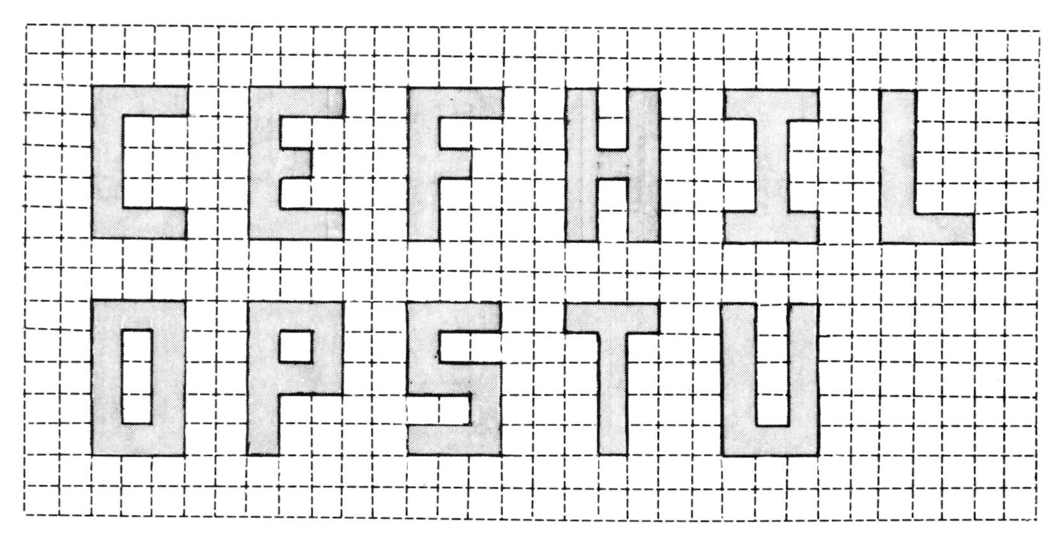

5. In the middle of a rectangular lawn measuring 32 m by 25 m, there is an ornamental pond 5 m long and 3 m wide. Find the area of turf on the lawn. Would it matter whereabouts the pond was situated? Would it matter if it were turned at an angle to the sides of the lawn?

6. Mr and Mrs Jones buy a rectangular plot of land measuring 45 m by 20 m. On it they build a house measuring 15 m by 10 m. They also make three rectangular flower-beds, one measuring 7 m by 2 m and two measuring 10 m by 1.5 m. The rest of the plot is covered with grass. Find the area of the grass. Does it matter where the house and flower-beds are situated?

7. An indoor swimming pool measures 25 m by 15 m. It is bordered by a surround 4 m wide, as in the diagram.
   Find (a) the area of the surface of the pool
   (b) the area of the 'room' it is in
   (c) the area of the surround.
   (d) If the surround is paved with tiles each measuring 10 cm by 10 cm, how many tiles are needed?

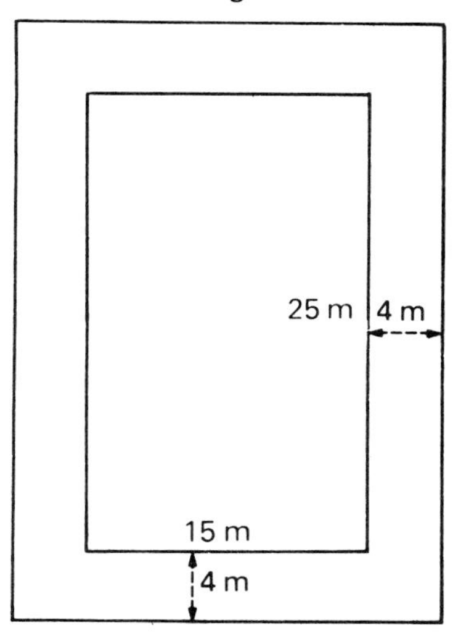

Exercise 8.4 *Mixed units*

Find the area of a rectangle with the given sides. Give your answer in the unit in brackets.

1.	10 cm, 1 m (cm^2)	**2.**	10 m, 0.5 km (m^2)	**3.**	2 cm, 2 m (cm^2)
4.	5 mm, 12 cm (cm^2)	**5.**	80 cm, 5 m (m^2)	**6.**	450 m, 8 km (km^2)
7.	9 cm, 45 m (m^2)	**8.**	15 mm, 200 m (m^2)	**9.**	5 mm, 3 m (cm^2)
10.	25 cm, 1 km (m^2)	**11.**	36 cm, 0.75 km (m^2)	**12.**	13 mm, 3 km (m^2)

13. A crêpe bandage is 6 cm wide and 5 m long. Find its area in m^2.

14. A roll contains sticking tape 75 m long and 1.2 cm wide. Find its area in m^2.

15. A tape measure has tape 1 cm wide and 50 m long. Find its area in m^2.

16. A room measures 6 m by 5 m. A skirting board 8 cm high runs right round the room, apart from two doorways each 1 m wide. Find the area of the skirting board.

17. The tape in a cassette is 4 mm wide and 30 m long. Find its area in cm^2.

18. A road is 15 m wide and 60 km long. Find its area in km^2.

9

Averages

Exercise 9.1 *Mean of whole numbers*

Find the average (mean) of each of the following sets of numbers.
Check that your answer looks sensible.

1. 6, 5, 10, 3
2. 12, 5, 27, 16, 1, 19, 4
3. 7, 42, 2, 63, 16
4. 3, 2, 19, 23, 13
5. 2, 17, 9, 12, 20, 14, 6, 16
6. 120, 47, 123, 31, 29
7. 107, 2, 124, 17, 132, 98
8. 72, 0, 3, 70, 100
9. 1, 3, 5, 7, 9
10. 10, 11, 12, 13, 14
11. 100, 110, 120, 130, 140
12. 110, 111, 112, 113, 114
13. 11, 13, 15, 17, 19
14. 4, 12, 20, 28, 36
15. 0, 0, 0, 0, 0, 0, 3, 5
16. 7, 4, 1, −1, −4, −7
17. 14, 11, 8, 6, 3, 0
18. 1, 11, 111, 1111, 11 111

Exercise 9.2 *Mean of fractions and decimals*

Find the average (mean) of each of the following sets of numbers.
Check that your answer looks sensible.

1. 8, 5, 7, 6
2. 17, 26, 3, 33, 13
3. 1.6, 13.2, 2.7, 5.1, 8.4
4. 12.0, 4.7, 12.3, 3.1, 2.9
5. $2\frac{1}{4}$, $1\frac{1}{3}$, $\frac{1}{2}$, $1\frac{1}{24}$, $2\frac{1}{8}$
6. 10.6, 3.5, 1.8, 16.2
7. $\frac{1}{2}$, $\frac{2}{3}$, $\frac{5}{6}$, $\frac{4}{5}$
8. 129, 8, 47, 33, 109, 17, 56

Exercise 9.3 *Mean of quantities*

Find the average of each of the following sets.
(Remember that 10:1 yr means 10 years and **one** month, not 10 years and $\frac{1}{10}$ year.)

1. 10:1 yr, 9:4 yr, 9:11 yr, 10:0 yr, 11:1 yr
2. 5.4 kg, 0.5 kg, 3.3 kg, 7.8 kg, 2.5 kg, 5.1 kg
3. 4.56 m, 5.39 m, 5.22 m, 4.61 m, 4.24 m, 5.74 m
4. 12:6 yr, 11:9 yr, 13:1 yr, 13:2 yr, 12:10 yr, 13:6 yr, 11:10 yr
5. 1.2 metres, 640 cm, 0.72 metres, 962 cm, 0.83 metres.
Answer in (**a**) metres, (**b**) cm.
6. 0.2 cm, 1.7 cm, 0.9 cm, 1.2 cm, 2.0 cm, 1.4 cm, 0.6 cm, 1.6 cm
7. 641 mg, 10.6 g, 1.8 kg (Answer in grams.)
8. $13\frac{1}{2}$ yr, $12\frac{1}{2}$ yr, 16 yr, $14\frac{1}{2}$ yr, 13 yr, $12\frac{1}{2}$ yr
9. £1.06, 32p, 74p, £1.26, 5p, 35p (Answer in pence.)
10. 12:2 yr, 14 yr, $13\frac{1}{2}$ yr, 12:8 yr, 11:7 yr, 11:11 yr, 12:3 yr

Exercise 9.4 *Missing numbers*

Find the value of x in each case. The average for the group is given in brackets.

1. 6, 5, 10, x (Av. 6)
2. 35, 16, 23, 41, x (Av. 26)
3. 112, 119, 131, 126, 108, 121, x (Av. 111)
4. 16, 21, 13, x (Av. 14.5)
5. 11:2 yr, 12:3 yr, 11:7 yr, 11:6 yr, 12:1 yr, 11:5 yr, x (Av. 11:9 yr)
6. 62, 14, 98, 36, 70, x (Av. 56)
7. 6.2, 3.6, 4.8, 9.2, 5.4, 7.3, x (Av. 5.6)
8. 0.1, 0.3, 0.7, 0.2, 0.5, x (Av. 0.47)
9. 10.1, 10.3, 10.1, 10.2, 10.3, 10.2, x (Av. 10.2)
10. 5, 1, 3, 4, 2, x (Av. 10)

Exercise 9.5 *Problems*

1. If the ages of three children are 10:1 yr, 8:11 yr and 9:6 yr respectively, what is their average age?
2. If the average length of five lines is 3.7 cm, and the average of the first four is 4.1 cm, what is the length of the fifth line?
3. A batsman scores an average of 34.25 runs in the first four innings of the season. In the next six innings he is out for 63, 0, 112, 76, 2, 7. What is his average for the ten innings?
4. A thermometer reads 64°, 58°, 62°, 60°, 65°, 57° at 11.00 a.m. on six successive days. What must the thermometer read on the seventh day to make the average for the week 60°?
5. A 50 kg bag of coal costs £4.20 if it is grade A and £3.70 if it is grade B. What would be the average cost per bag of an order of 6 bags of grade A and 4 bags of grade B?
6. A man finds a clay tablet of the year 1200 BC and is just able to decipher the inscription 'The total rainfall for 11 months was 63.2 drops, although the rainfall for the year averaged 76.2 drips per month.' How many drips fell in the twelfth month? (He had already discovered that there were 10 drips to a drop and that the number system used was denary.)
7. The average mark of the top three members of a form was 63%, the average for the bottom four was 24%. If the overall average of the form of fifteen was 43%, what was the average of the remaining eight members?
8. A pupil has an average mark of 65% for English and Maths, and an average mark of 41% for French and German. What must his average mark have been in Geography and History if his combined average was 50% for all six subjects?

10

Distance, speed and time

Exercise 10.1 *Simple calculations*

What are the following times in minutes as fractions of an hour?

1.	30	**2.**	20	**3.**	15	**4.**	12	**5.**	5
6.	6	**7.**	36	**8.**	35	**9.**	50	**10.**	27

What are the following fractions of an hour expressed in minutes?

11.	$\frac{1}{4}$	**12.**	$\frac{1}{6}$	**13.**	$\frac{1}{30}$	**14.**	$\frac{5}{12}$	**15.**	$\frac{2}{3}$
16.	$\frac{4}{5}$	**17.**	$\frac{7}{12}$	**18.**	$\frac{11}{20}$	**19.**	$\frac{5}{8}$	**20.**	$\frac{1}{36}$

21. How far would a car travel if it travelled at 60 km/h for the following times?
(**a**) 1 hour (**b**) 30 min (**c**) 6 min (**d**) $2\frac{1}{2}$ hours (**e**) 37 min

22. How far would a car travel if it travelled at 40 km/h for the following times?
(**a**) $1\frac{1}{2}$ hours (**b**) 6 min (**c**) 24 min (**d**) $4\frac{1}{4}$ hours (**e**) 50 min

23. How long would it take a car travelling at 60 km/h to travel these distances?
(**a**) 120 km (**b**) 30 km (**c**) 12 km (**d**) 90 km (**e**) 27 km

24. How long would it take a car travelling at 45 km/h to travel these distances?
(**a**) 135 km (**b**) 30 km (**c**) 27 km (**d**) 42 km (**e**) 141 km

25. What is the speed in km/h of a car travelling 72 km in the following times?
(**a**) 2 hours (**b**) $2\frac{1}{2}$ hours (**c**) 36 min (**d**) 54 min (**e**) $1\frac{1}{5}$ hours

26. What is the speed in km/h of a car travelling the following distances in $1\frac{1}{2}$ hours?
(**a**) 60 km (**b**) 72 km (**c**) 105 km (**d**) 66 km (**e**) 93 km

27. Replace the question marks in the following table with the appropriate answers:

	Dist. (km)	Speed (km/h)	Time (h)
(a)	32	16	?
(b)	40	?	2
(c)	?	30	$1\frac{1}{2}$
(d)	?	$37\frac{1}{2}$	4
(e)	72	?	$2\frac{4}{7}$
(f)	108	30	?

28. Replace the question marks in the following table with the appropriate answers:

	Dist.	Speed	Time
(a)	500 m	? m/s	10 s
(b)	1 km	? m/s	50 s
(c)	64 km	20 km/h	? h
(d)	700 km	28 km/h	? h
(e)	? km	5 m/s	5 h
(f)	? m	30 km/h	20 min

Exercise 10.2 *Further calculations*

1. A man cycles for 30 km at 20 km/h. How long does the journey take?
2. A car is driven at an average speed of 53 km/h for 2 h. How far does it travel?
3. An athlete ran 800 metres in 1 minute 40 seconds. What was his average speed in m/s?
4. A car travels 63 km in $1\frac{3}{4}$ hours. What is its average speed in km/h?
5. A motorist travels for $2\frac{1}{2}$ hours at 72 km/h. What is the distance travelled?
6. What would be the average speed of a car travelling 120 km in $2\frac{2}{5}$ h?
7. How far would a car go in 12 minutes at 55 km/h?
8. How far would a car go in 5 minutes at 48 km/h?
9. 12.5 km in 10 minutes is what speed in km/h?
10. 15 km in 24 minutes is what speed in km/h?
11. 42 km at 70 km/h takes how many minutes?
12. 100 metres in 10 seconds is what speed in km/h?

Exercise 10.3 *Problems*

1. A car goes at 60 km/h for 3 hours and then at 80 km/h for 2 hours.
(a) What is the total distance travelled?
(b) What is the total time taken?
(c) What is the average speed for the whole journey?

2. A car goes for $2\frac{1}{2}$ hours at 72 km/h and then for $4\frac{1}{2}$ hours at 86 km/h.
(a) What is the total time taken?
(b) What is the total distance travelled?
(c) What is the average speed for the whole journey?

3. A car goes for $3\frac{1}{4}$ hours at 120 km/h and then for $4\frac{1}{4}$ hours at 48 km/h. What is the average speed for the whole journey?

4. A car goes for $\frac{1}{4}$ h at 80 km/h and then for $\frac{3}{4}$ h at 60 km/h. Find the average speed for the whole journey.

5. A car goes 20 km at 60 km/h and then 10 km at 45 km/h. How many minutes does this take altogether?

6. A motorist averages 75 km/h for 1 hour 12 minutes. How far will she travel? At what speed should she travel in the next 48 minutes to raise her average speed to 77 km/h for the whole journey?

7. On a journey a motorist travels 144 km in 2 hours. At what speed should he travel back if his average speed for both parts together is to be 80 km/h?

8. Motorist A travels from town X to town Y at 72 km/h. Motorist B travels to town X from town Y, starting at the same time, at 96 km/h. The distance between X and Y is 140 km.
   (a) If they meet after t hours, how far has A travelled (in terms of t)?
   (b) If they meet after t hours, how far has B travelled (in terms of t)?
   (c) What is the ratio of their speeds?
   (d) What is the ratio of the distances they have travelled when they meet?
   (e) Use these ideas to find how far each will have travelled when they meet.

9. Cyclist A sets out from Boonville at 1200 and travels at 20 km/h. At 1300 cyclist B sets out from Boonville along the same route as A and travels at 30 km/h.
   (a) How far apart are they at 1300?
   (b) What is the ratio of their speeds after 1300?
   (c) What is the ratio of the distances they travel between 1300 and any later time?
   (d) When B overtakes A, the difference between these distances will be the answer to (a). Use this to find how far from Boonville and at what time B overtakes A.

11

Statistical graphs

Exercise 11.1 *Column graphs*

1. The column graph on the right shows the results obtained by five friends in a test out of ten.
   (**a**) Who had the lowest mark?
   (**b**) What was the highest mark?
   (**c**) How many marks did Andrew get?
   (**d**) By how many marks did Elizabeth beat Anne?

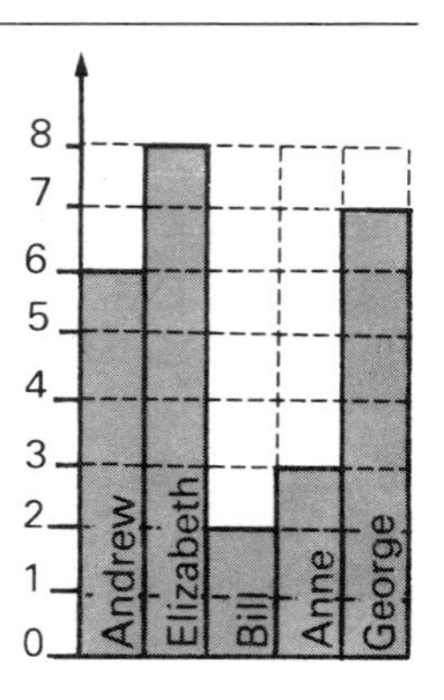

2. In a later test the five friends had the following marks:

Andrew	5
Elizabeth	6
Bill	8
Anne	7
George	3

   Draw a column graph like that in Question 1 to illustrate these results.

3. A class of 30 children were asked 'What is your favourite pet?' with the following results:

Cats	8
Dogs	14
Rabbits	3
Mice	1
Others	4

   Draw a column graph to illustrate this information.

4. The column graph on the right shows the daily rainfall, measured in millimetres, for one week.
   (**a**) On what day was the greatest rainfall?
   (**b**) How much rainfall was there on Saturday?
   (**c**) What was the total rainfall for the week?
   (**d**) What was the average rainfall per day?

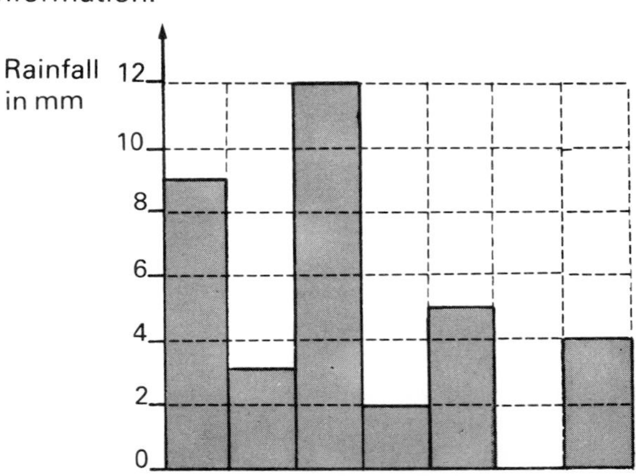

5. The column graph below shows the number of cars per hour passing the school gate, from 0900 hours to 1500 hours. (For simple calculations the totals have been rounded to the nearest ten).
   (a) How many cars passed during each hour?
   (b) What was the total number of cars passing during the whole time?
   (c) What was the average number of cars passing per hour?
   (d) What do you think accounted for the large number of cars during the periods 0900 – 1000 and 1600 – 1700?

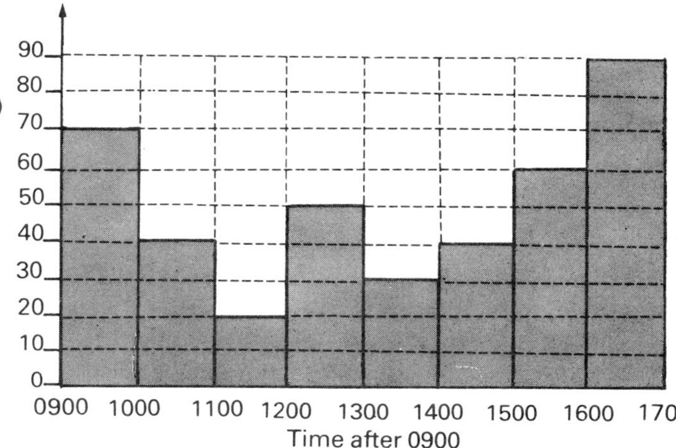

6. The column graph on the right shows the results of a school football team. The number of matches at which they scored a certain number of goals has been recorded.
   (a) What does the first column show?
   (b) How many matches are represented in the column in which one goal was scored?
   (c) How many matches were played altogether?
   (d) What was the total number of goals scored?

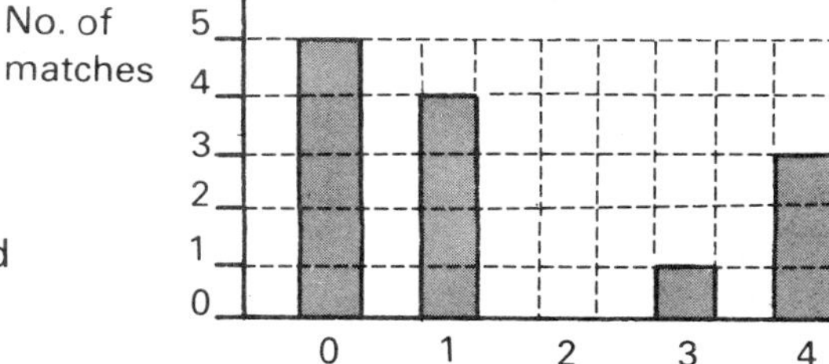

7. The rainfall for a certain town was as follows. (All measurements are in millimetres, rounded to the nearest five.)

Jan	Feb	Mar	Apr	May	Jun	Jul	Aug	Sep	Oct	Nov	Dec
45	30	60	65	30	20	10	15	5	45	65	60

   (a) Plot a column graph to illustrate these figures, taking a vertical scale of 1 cm to 5 mm of rainfall.
   (b) Calculate the average rainfall per month. Draw this line on your graph.
   (c) Calculate how much each month's rainfall is above or below the mean (average). What do you notice about the totals above and below the mean?

8. The results for a test out of 49 marks for Form 6Z are shown on the column graph on the right.
   (a) How many pupils are there in the class?
   (b) What is the largest possible total mark for the whole form?
   (c) What is the smallest possible total mark for the whole form?

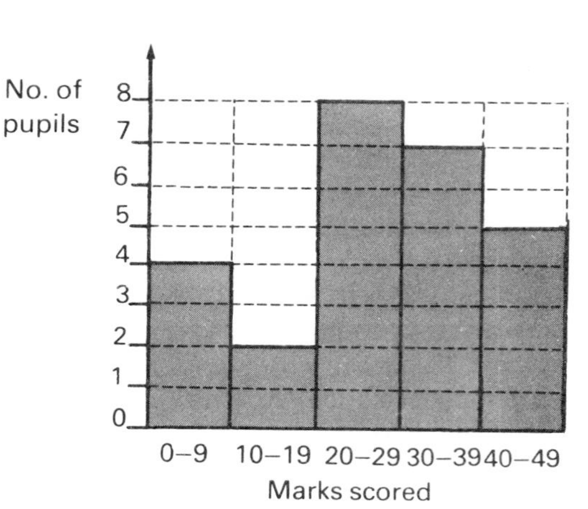

Exercise 11.2 *Pie charts*

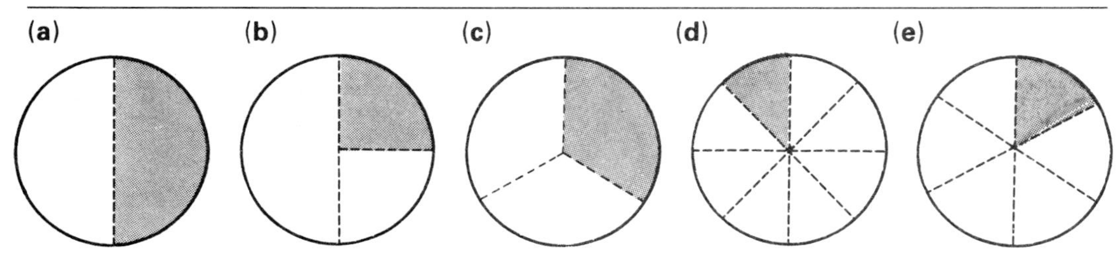

1. What fraction of the circle is shaded in each of circles (a)–(e)?
2. What is the angle at the centre of the circle for each shaded section in (a)–(e)?
3. What are each of these angles as fractions of 360°?
   (Cancel down if possible.)

(**Note**: A section cut from the centre of a circle in this way is called a sector.)

Express the following angles as fractions of 360° in their lowest terms:

4. 90°	**5.** 60°	**6.** 270°	**7.** 180°
8. 36°	**9.** 108°	**10.** 45°	**11.** 40°

What fraction of the circle should be shaded to represent the following angles?

12. 60°	**13.** 180°	**14.** 270°	**15.** 135°
16. 300°	**17.** 144°	**18.** 40°	**19.** 330°

What angles would represent the following fractions of a circle?

20. $\frac{1}{4}$	**21.** $\frac{2}{3}$	**22.** $\frac{1}{5}$	**23.** $\frac{1}{8}$
24. $\frac{1}{9}$	**25.** $\frac{5}{6}$	**26.** $\frac{5}{12}$	**27.** $\frac{13}{18}$

If the following numbers of children were to be represented on a pie chart, what angle would represent one child?

28. 360	**29.** 180	**30.** 120	**31.** 90
32. 72	**33.** 720	**34.** 364*	**35.** 124*

(*Think carefully about how accurately the pie chart would be read.)

36. The circle opposite represents a school of 360 pupils. The shaded sector represents the pupils with fair hair.

(**a**) How many degrees represent one pupil?

(**b**) How many pupils have fair hair?

(**c**) What fraction of the school have fair hair?

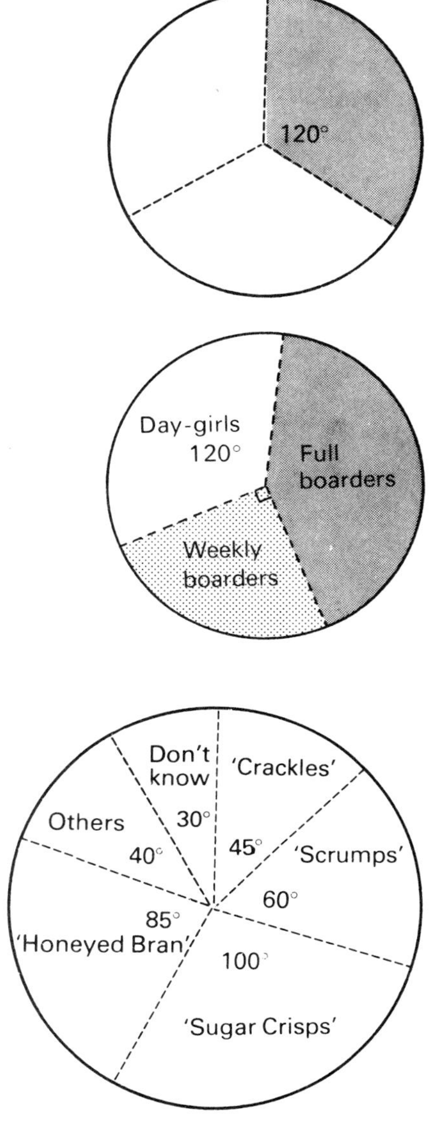

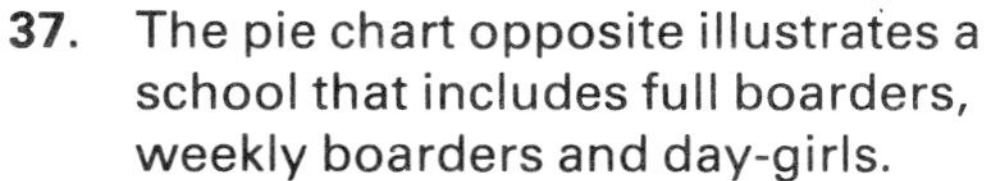

37. The pie chart opposite illustrates a school that includes full boarders, weekly boarders and day-girls.

(**a**) What fraction of the school are day-girls?

(**b**) What fraction of the school are weekly boarders?

(**c**) If the number of girls in the school is 300, how many degrees represent 5 girls?

(**d**) How many girls are full boarders?

38. The pie chart on the right illustrates the answers of 72 pupils in a school to the question 'Which kind of cereal do you prefer to have at breakfast?'

(**a**) How many degrees represent one pupil?

(**b**) Which cereal proved to be the most popular?

(**c**) How many pupils voted for 'Sugar Crisps'?

(**d**) What fraction of the group voted for 'Honeyed Bran'?

(**e**) Which two would cater for the tastes of more than half of the pupils?

39. Out of 180 people, 90 have brown eyes and 40 have blue eyes.

(**a**) What fraction of the people have brown eyes?

(**b**) What fraction of the people have blue eyes?

(**c**) Show these results by dividing up a circle into sectors to make a pie chart. (Label each sector clearly and mark the angles at the centre of the circle.)

40. Out of a form of 18 pupils 4 have dogs as pets, 9 have cats and the remainder have no pets. You are going to draw a pie chart to show this information.

(**a**) What angle would represent one pupil?

(**b**) What angle would represent four pupils?

(**c**) What angle would represent nine pupils?

(**d**) Show this information on a pie chart, labelling the sectors clearly.

41. In a class of 36 pupils 12 travel to school entirely by bus, 3 walk to school and 9 travel entirely by train. The remainder are boarders.

(**a**) What fraction of the form travel (**i**) by bus (**ii**) by train (**iii**) on foot?

(**b**) On a pie chart, what angle would represent one pupil?

(**c**) Using a circle of at least 4 cm radius, draw a pie chart of the original information. Make sure that you label each sector clearly. Mark the angle at the centre clearly in each case.

42. In the 24 hours of a day a boy worked out that he spent 10 hours sleeping, 2 hours eating, 7 hours at school, 2 hours watching television and 3 hours on other activities. Illustrate this information with a pie chart.

(**a**) What angle will represent one hour?

(**b**) Work out the angles at the centre for each sector and mark them in clearly, also labelling the sectors clearly.

(**c**) Work out a scheme for a day in the holidays in order to be able to compare it with a day in term time.

12

Introduction to algebra

Exercise 12.1 *Elementary use of letters*

Discuss the following questions:

1. (a) What number is three more than 5?
   (b) Think how you arrived at your result for (a) and then give the number that is three more than x.
2. (a) What number is 7 more than 12?
   (b) Think how you arrived at your result for (a) and then give the number that is x more than 12.
3. (a) What number is 2 less than 7?
   (b) Think how you arrived at your result for (a) and then give the number that is two less than x.
4. (a) What number is 6 less than 20?
   (b) Think how you arrived at your result for (a) and then give the number that is x less than 20.

Using the ideas in Questions 1–4, answer the following:

5. What number is 2 bigger than y?
6. What number is 5 smaller than n?
7. What number is t more than 4?
8. What number is z less than 14?
9. What number is 3 greater than y?
10. What number is n less than 14?
11. Describe what is meant by $d + 3$.
12. Describe what is meant by $g - 4$.
13. Describe what is meant by $8 + z$.
14. Describe what is meant by $9 - v$.
15. Describe what is meant by $x - y$.

Exercise 12.2 *Further use of letters*

1. If I walk for x kilometres and run for 2 kilometres how far have I gone altogether?
2. A post is 8 feet long. If 6 feet is painted, what length remains to be painted?
3. A post is x feet long. If 5 feet is painted, what length remains to be painted?
4. A post is 7 feet long. x feet is painted. What length remains to be painted?
5. If $2n$ means $2 \times n$, what is the meaning of $5n$?
6. If one orange costs 5 pence, what would be the cost of 5 oranges? (Think how you work this out.)
7. If one orange costs 5 pence, what would be the cost of x oranges?
8. If one orange costs x pence, what would be the cost of 7 oranges?

9. If I have 60 pence and spend x pence how much have I left?
10. If I have y pence and spend 23 pence how much have I left?
11. By how much is 17 bigger than 14? (Think how you work this out.)
12. By how much is g bigger than 14?
13. By how much is 17 bigger than j?
14. What number is 3 times as big as j?
15. How many pence are there in £5? (Think how you work this out.)
16. How many pence are there in £p?
17. How many millimetres are there in 3 centimetres? (Think how you work this out.)
18. How many millimetres are there in b centimetres?
19. Divide 16 sweets between 4 children equally. How many will each receive? (Think how you work this out.)
20. Divide s sweets between 4 children equally. How many will each receive?
21. Divide 16 sweets between c children equally. How many will each receive?
22. What number is one more than c?
23. What number is one less than c?
24. Write down three consecutive numbers of which c is the middle number.
25. What number is one smaller than twice b?
26. What number is 4 greater than three times a?
27. Write down three consecutive numbers of which c is the smallest.
28. Write down three consecutive numbers of which c is the largest.
29. If n is any even number, what would be the next higher even number?
30. If f families have 6 children each, how many children are there altogether?
31. If r times s is written rs what is the meaning of xy?
32. If f families each have c children, how many children are there altogether?
33. If n is any number, write down an expression with n in it that is
    (**a**) even (**b**) odd.
34. If a and b are any two numbers, what is their average?
35. By how much is 7 greater than 2? (Think how you work this out.)
36. By how much is r greater than 2?
37. By how much is 7 greater than r?
38. By how much is r greater than t?
39. By how much is y less than 13?
40. By how much is 13 less than y?

Exercise 12.3 *Letters in equations*

Discuss the following questions:

1. What number when made 5 bigger becomes 13?
2. What number when 4 is added becomes 10?
3. What number when made 5 smaller becomes 19?
4. What number when 7 is subtracted from it becomes 2?
5. What number when increased by 8 becomes 21?
6. What number when decreased by 6 becomes 11?
7. What number when added to itself becomes 14?
8. What number when doubled becomes 18?

9. When 4 is added to 3 times a number the result is 16.
   (a) What would be 3 times the number?
   (b) What is the number?

Find the numbers represented by the letters in the following equations:

10. $x+5=13$ **11.** $a+4=10$ **12.** $b+6=8$
13. $a+3=15$ **14.** $x-5=19$ **15.** $y+8=21$
16. $h-7=2$ **17.** $d+2=3$ **18.** $j+8=21$
19. $y-6=11$ **20.** $x+x=14$ **21.** $4+r=10$
22. $2x+1=17$ **23.** $15=6+b$ **24.** $7=x-3$
25. $3=n-2$ **26.** $3d=12$ **27.** $3d+2=17$
28. $2x-4=14$ **29.** $n-2=0$

Exercise 12.4 *Directed numbers*

Make a copy of this number line in your books and use it to find the answers to the questions below.

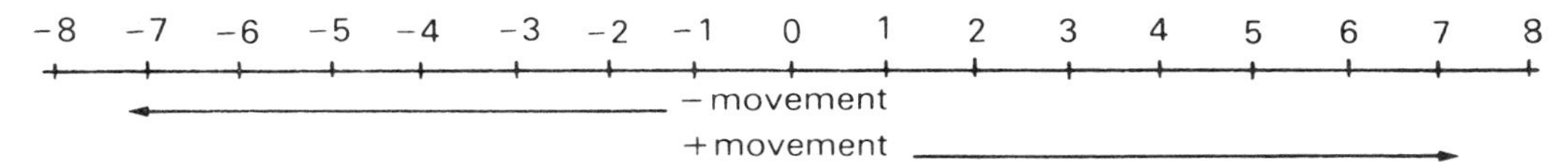

Replace the question marks in the following table with the appropriate answers:

	Starting point	Movement	Arrival point
Example:	+2	+6	? (+8)
1.	+5	−3	?
2.	−2	+7	?
3.	−7	+3	?
4.	−5	+5	?
5.	+3	−8	?
6.	−1	?	−6
7.	+3	?	+8
8.	0	?	−4
9.	?	+7	+1
10.	?	−2	+5
11.	?	−7	−2
12.	?	+7	+2
13.	−4	+2	?
14.	−1	?	−5

Using the ideas in Questions 1–14, find the value of each of the following:

15. $+2-1$ **16.** $-3+6$ **17.** $0-5$
18. $-1-4$ **19.** $-4-1$ **20.** $+8-10$
21. $-8+10$ **22.** $-7+4$ **23.** $-1+7$
24. $+5-0$ **25.** $-2-2-2$ **26.** $0-4+6$
27. $+5-6+3$ **28.** $-2+7-3$ **29.** $-4+0-4$

Find the value of each of the following:

30. $9-6$
31. $6-9$
32. $3-8$
33. $8-3$
34. $2-12$
35. $-5-9$
36. $-3-2+5$
37. $6-11+7$
38. $-6+6+3$
39. $-5-4+9$
40. $7-10+10$
41. $-4+6-3+2$
42. $10-4+7-11$
43. $8-14$
44. $29-18$

In each of the following pairs of numbers, which number is the greater and by how much?

45. $+5,\ +2$
46. $+2,\ -4$
47. $0,\ -2$
48. $-1,\ -7$
49. $-2,\ +6$
50. $+4,\ 0$
51. $-3,\ +2$
52. $-12,\ -8$
53. $+12,\ +8$
54. $+20,\ -16$
55. $9,\ +9$
56. $-7,\ -9$
57. $-1,\ +2$
58. $+1,\ -2$
59. $-3,\ +2$

Exercise 12.5 *Rules of signs*

For this exercise you will need to make a copy of the number line used in Exercise 12.4
It will be helpful if the minus sign is regarded as a 'reverser',
i.e. -4 is in the reverse direction to $+4$.

Starting at zero each time, where would you arrive if you made the following moves?

1. -3 followed by another -3
2. two 'lots' of -3
3. 2×-3
4. $+2\times -3$
5. $-3\times +2$ (Think carefully)

Starting at zero each time, where would you arrive if you made the following moves?

6. Three 'lots' of -2
7. 3×-2
8. $+3\times -2$
9. $-(+3\times -2)$ (Remember 'minus' as a reversing factor.)
10. Can you work out where you would arrive if you moved -3×-2?
(How does this connect with the previous question?)

Starting at zero each time, where would you arrive if you made the following moves?

11. 2×-3
12. 3×-2
13. -3×-2
14. 4×2
15. -4×-2
16. 5×-1
17. 0×-4
18. -5×-1
19. $+4\times -2$
20. $-4\times +2$
21. -1×-2
22. $+1\times -2$
23. $-2\times +5$
24. -5×-2
25. -3×-1
26. -3×0
27. $+1\times -5$
28. -1×-5
29. -6×-3
30. $+6\times -3$
31. How many 'lots' of -2 are there in -6? (Look at your answer to Question 12.)
32. What is the answer to -6 divided by -2?
33. What multiplied by -2 makes -6? (Think carefully!)
34. What is the answer to -6 divided by $+3$?
35. What multiplied by -3 makes $+6$?
36. What is the answer to $+6$ divided by -3?

Find the value of each of the following:

37. $6 \div 2$
38. $+6 \div +2$
39. $-6 \div -2$
40. $-6 \div +2$
41. $-8 \div +4$
42. $-8 \div -4$
43. $-4 \times +2$
44. $5 \div 2$
45. $+5 \div -2$
46. $-5 \div +2$
47. $-10 \div +5$
48. $+10 \div -5$
49. $+10 \div +5$
50. $-7 \div -2$
51. $-4 \div +1$

Find the value of each of the following:

52. $+3 \times +3$
53. -3×-3
54. $(-3)^2$
55. $-2 \times -2 \times -2$
56. $(-2)^3$
57. $(2)^3$
58. $(-2)^2 \times -1$
59. $(3)^2 \times -1$
60. The square of -4
61. The square of $+3$
62. The square root of $+9$ (careful!)
63. The square root of $+16$

Exercise 12.6 *Substitution – whole numbers*

Find the value of each of the following if $a=6$, $b=4$, $c=2$ and $d=0$:

1. $a+b$
2. $b+c$
3. $a-b$
4. $b-c$
5. $c+d$
6. $b-d$
7. $a+d$
8. $b-a$
9. $b-a+c$
10. $d-c$
11. $d-b$
12. $b+c+d$
13. $a+b-c-d$
14. $b-a+c-d$
15. $b+c-a-d$
16. $c+d-a+b$
17. $b-a+d+c$
18. $a-b+c-d$
19. $a-b-c-d$
20. $b-a+d-c$

Find the value of each of the following if $a=2$, $b=3$, $c=4$ and $d=-1$:

21. $3a$
22. $2c$
23. $7b$
24. ab
25. bc
26. $4d$
27. $2c-3$
28. $2c-3a$
29. $3a-c$
30. $5a+d$
31. $2c+2d$
32. $a+2d$
33. $4d-a$
34. $3a-d$
35. b^2
36. d^2
37. a^2-b^2
38. a^2+d^2
39. b^2-a^2
40. a^2-d^2

If $a=12$, $b=3$, $c=-2$ and $d=0$, find the value of each of the following:

41. $a+c$
42. $b+c$
43. $c+d$
44. $a-b$
45. $a-c$
46. $c-d$
47. $c+c$
48. $2c$
49. bd
50. ac
51. abc
52. abd
53. acd
54. $ac+ac$
55. $ab-ac$
56. $ad+bc$
57. $ad-bc$
58. $ad-bd$
59. $2b+3c+4c$
60. $a-4b$

If $a=12$, $b=3$, $c=-2$, and $d=0$, find the value of each of the following:

61. $a \div b$
62. $a \div c$
63. $ab \div b$
64. $ac \div b$
65. $ab \div c$
66. $cd \div a$
67. $a \div bc$
68. $\dfrac{a+d}{b}$
69. $\dfrac{a+3c}{b}$
70. $\dfrac{a+100d}{b}$
71. $\dfrac{a-cd}{b}$
72. $\dfrac{d-c}{c}$
73. $\dfrac{2b-3c}{b}$
74. cd^5
75. c^2+2b
76. $\dfrac{b^2-c^2}{a}$

Exercise 12.7 *Collecting like terms*

Add together:

1.	$3a$ and a	**2.**	$7b$ and $5b$	**3.**	$3c$ and $7c$
4.	$10ab$ and $5ab$	**5.**	$-x$ and $-4x$	**6.**	$4x^2$ and $3x^2$
7.	$-5y$ and $-11y$	**8.**	$-3z$ and $-5z$	**9.**	$-2y^2$ and $-4y^2$
10.	$7c^2$ and $-4c^2$	**11.**	a, $6a$ and $-3a$	**12.**	$9x, -3x, x$
13.	$10b, -5b, -2b$	**14.**	$17xpq$ and $2xpq$	**15.**	$-a^2$, $3a^2$ and $-2a^2$

Simplify each of the following:

16.	$10b - b - 7b + 4b$	**17.**	$-5c + 12c - 4c + c$
18.	$8p - 11p + 5p - p$	**19.**	$9c + 2c - 3c + 6c$
20.	$-a^2 + 12a^2 - 6a^2 + 10a^2$	**21.**	$-5xy + 6xy + 7xy - 4xy$
22.	$9ab - 12ab + 5ab - 2ab$	**23.**	$-7a^2b + 3a^2b - 4a^2b + 6a^2b$
24.	$12a^3 - 5a^3 + 2a^3$	**25.**	$an^2 - 3an^2 + 2an^2 - an^2$

Collect up like terms in the following:

26.	$a + 2x - 3x + 2a$	**27.**	$3x + 2y - x$	**28.**	$5x - 3y + 2x$
29.	$2a + 3b - b - a$	**30.**	$5a - 7b + 8a - 13b$	**31.**	$x^2 + x - 2x + xy$
32.	$3r - 4rs + 2s$	**33.**	$2y - 3z - y + z$	**34.**	$4t - 2p + t + 4p$
35.	$3t - 4b - t + 2b$	**36.**	$3a - 2b + b - 2a$	**37.**	$x^2 + x + x^2 - x$
38.	$a - b + b - a$	**39.**	$a + b + a - b$	**40.**	$a + b - a - b$
41.	$2x - 6 - x + 5$	**42.**	$1 - x + x - x^2$	**43.**	$2x - 4 + 3 - x$
44.	$x^2 - x - 2x - x^2$	**45.**	$8x^2 - 3x - 7x + 16$	**46.**	$2ab + 4ba - 3ba$
47.	$2xy + 3yx - 2yz$	**48.**	$2xy + y - yx - x$	**49.**	$7xy - 2ab + xy - 5ab$

Exercise 12.8 *Further equations – negative and fractional solutions*

1. If a number were 4 bigger it would be 1. What is the number?
2. What number when increased by 7 becomes -1?
3. When a certain number is subtracted from 3 the result is 6. What was the number subtracted?
4. When 4 is subtracted from three times a number the result is -19. What was the number?
5. When 3 is added to twice a certain number the result is 8. What was the number?

Find the value of x in the following equations:

6.	$x + 3 = 2$	**7.**	$2x + 1 = 6$
8.	$3 - x = 5$	**9.**	$7 = x + 2$
10.	$7 = x + 12$	**11.**	$5x + 1 = 7$
12.	$2x + 4 = -1$	**13.**	$3x - 4 = -7$
14.	$x + 7 = 3x$	**15.**	$2x + 1 = x + 2$

Solve the following equations:

16.	$3n = 15$	**17.**	$4x = 12$	**18.**	$6y = 24$	**19.**	$7x = 21$
20.	$9c = 63$	**21.**	$4n = 13$	**22.**	$2g = 5$	**23.**	$6b = 14$
24.	$7f = 5$	**25.**	$5x = -10$	**26.**	$-3n = 12$	**27.**	$-6x = -12$

28. $-5z=-11$ 29. $6v=-13$ 30. $7g=-21$ 31. $-7d=-21$
32. $7d=21$ 33. $13a=91$ 34. $-7a=91$ 35. $3r=-2$

Solve the following equations:

36. $3n+2=17$ 37. $6h-2=22$ 38. $4v+1=14$
39. $7t-5=16$ 40. $5x+10=-20$ 41. $2-3n=14$
42. $1-6x=-13$ 43. $4-5x=-15$ 44. $7t+6=-15$
45. $3w+2=0$ 46. $3-5d=-8$ 47. $13x-9=100$
48. $3q+12=0$ 49. $8+7t=29$ 50. $6f+8=-5$

Solve the following equations:

51. $0.3x=1.2$ 52. $1.2b=0.6$ 53. $1.2c=6$
54. $0.5t=1.6$ 55. $0.2y=3.2$ 56. $0.4t=2.0$
57. $1.5b=6$ 58. $2.5v=7.5$ 59. $1.2f=4.8$
60. $7.2c=3.6$ 61. $6n=3.6$ 62. $4g=1.6$
63. $0.4h=0.16$ 64. $1.4s=4.2$ 65. $0.6t=2.4$

66. For what value of x is $4x$ the same as 24?
67. For what value of n is $5n$ the same as 45?
68. For what value of c is $7c$ the same as 63?
69. For what value of t is $6t$ the same as 30?
70. For what value of h is $2h+1$ the same as 13?
71. For what value of g is $3g+1$ the same as 28?
72. For what value of f is $4f-2$ the same as 14?
73. For what value of x is $4x-3$ the same as 19?
74. For what value of y is $4+3y$ the same as 25?
75. For what value of h is $2h-7$ the same as 13?
76. For what value of d is $4d-2$ the same as 26?

Exercise 12.9 *Removing brackets*

Remove the brackets in each of the following:

1. $2(a+b)$ 2. $3(a-b)$ 3. $2(x-1)$ 4. $4(x+y)$
5. $7(2-y)$ 6. $4(2a-1)$ 7. $-3(2a+1)$ 8. $-5(a+2)$
9. $-7(z-k)$ 10. $-(1-2a)$ 11. $-5(a+2b)$ 12. $-3(4-ax)$
13. $a(b+2)$ 14. $d(a+3d)$ 15. $-2(4-2t)$ 16. $-c(4-2t)$
17. $2b(a+5)$ 18. $-c(-2+c)$ 19. $7(a-2b+c)$ 20. $7(a-2b)+c$

Remove the brackets in each of the following, and collect like terms where possible:

21. $2a+3(a+b)$ 22. $4c+2(3d-c)$
23. $5x-3(x+2y)$ 24. $8x-6y+4(2x-3y)$
25. $7f-2(d+3f)$ 26. $2(x+y)+3(x-y)$
27. $3(x-y)-2(x-y)$ 28. $7(a+b)-2a+3b$
29. $7(a+b)-2(a+3b)$ 30. $3x+2(x+y)$
31. $3x-2(x+y)$ 32. $3(x-2)+y$
33. $3x-2(x-y)$ 34. $3(a+b)-(2a+b)$
35. $2(x-y)-3(x-y)$ 36. $7(x-y)+2(x+y)$
37. $a(2+a)-3(2-a^2)$ 38. $7(x+2y)-4(x-y)$

Add:

39. $2a+1$ to $3a$

40. $2a-3$ to $3a+5$

41. $5a$ to $3-3a$

42. $a-1$ to $a+6$

43. $3a-5$ to $5-3a$

44. $-2c-7b$ to $c-5b$

45. $5a-3y$ to $-4a+5y$

46. $2z+5$ to $3-2z$

47. $7x+3y$ to $x-5y$

48. $-17x+23$ to $20x-17$

Subtract:

49. $3a$ from $4a+1$

50. $3a+2$ from $4a+1$

51. $3a-2$ from $4a+1$

52. $a-3$ from $2a+5$

53. $a+4b$ from $a+6b$

54. $a-2b$ from $b-3a$

55. $-4xy+5yz$ from $6yz+3xy$

56. $3x^2+y^2$ from $2x^2-3y^2$

57. $3c+2d$ from $5c$

58. $2a-3$ from $3a-2$

Exercise 12.10 *Introduction to formulae*

1. In a cricket eleven 3 players made a runs each, 4 made b runs each and the remaining players made c runs each. Assuming that there were no extras, what was the team's total score?

2. A pupil scores x marks in a test. If he loses y marks, what was the total mark for the test?

3. In a class x pupils made 5 mistakes each for a piece of work and y pupils made 6 mistakes each. What was the total number of mistakes made by the class?

4. A string 1 metre long is c centimetres too short to tie up a parcel. What was the length of the string required to tie up the parcel?

5. A string m metres long is c centimetres too short to tie up a parcel. What was the length of the string required to tie up the parcel?

6. A girl's stride is 60 cm. How far will she go in t strides?

7. A boy's stride is x cm. How far will he travel in t strides?

8. A sheet of 12p stamps contains x rows with y in each row. What would be the total cost of a sheet (**a**) in pence (**b**) in pounds?

9. If I give s sweets to p people and have h left over, how many did I have to begin with?

10. If the difference between two numbers is 5 and the smaller number is r, what is the larger in terms of r?

11. What is the sum of three consecutive numbers, the smallest being n?

12. What is the smallest number that is always divisible by both p and q exactly?

13. What is the sum of three consecutive numbers, the middle one being n?

14. What is the sum of five consecutive numbers, the middle one being h?

15. Tom starts a game of marbles with r marbles. Andy starts with q marbles. If Tom wins 7 from Andy, how many will each have now?

16. A boy has s sweets in each of three tins. If he takes 7 from one tin and 4 from each of the others, how many sweets are left in the tins altogether?

17. If I travel for 3 hours at x km/h how far will I travel?

18. If I travel for t hours at x km/h how far will I travel?

19. A train travels at a km/h for x hours and b km/h for y hours. How far will it travel altogether?

20. The length of a rectangular classroom is l metres and its width is w metres. What is the area of the floor of the classroom in square metres?

21. In the same classroom as Question 20, what would be the total area of the walls of the room if they are 3 metres high? (You may forget about any area taken up by the door or windows.)
22. If x apples cost y pence altogether, what would d similar apples cost?
23. A two-digit number has its units digit as y and the other digit as x. What is the total value of the number?
24. Two adjacent sides of a rectangle are $(a + b)$ cm and $(a - b)$ cm in length. What would be the perimeter of the rectangle?
25. A garden is m metres longer than it is wide. If its width is x metres what is the perimeter of the garden?
26. What is the perimeter of a rectangle that has a length of $(a - b)$ metres and a width of $(b + a)$ metres?
27. A girl has $(10x + 2)$ pence in her purse. She spends $2x$ pence on some stamps, and $(2 + x)$ pence on some toffee. How much is left in the purse?
28. A farmer goes to market with $(13g + 2)$ pounds in his wallet; he paid for 5 cows at g pounds each, and q sheep at 4 pounds each. How many pounds did he have left in his wallet?
29. If a train journey costs p pence for a journey of j kilometres, how much would a journey of k kilometres cost at the same rate?
30. A pen and a pencil cost $(3x + 3)$ pence altogether, while the pencil costs $(x + 4)$ pence. What would be the cost of a pen in pence?

Exercise 12.11 *Two-stage equations*

Find the value of x for which the following mathematical statements (equations) are true:

1. $3x + 7 = 2x + 9$
2. $4x + 3 = 2x + 7$
3. $2x + 10 = x + 15$
4. $5x + 6 = x + 10$
5. $7x + 1 = 2x + 16$
6. $4x + 6 = 3x + 9$
7. $5x + 2 = 2x + 5$
8. $6x + 1 = x + 4$
9. $2x + 3 = 3x + 1$
10. $5x + 2 = 2x - 4$

Solve the equations below for x:

11. $4x - 4 = x + 8$
12. $5x - 1 = 3x + 9$
13. $3x - 7 = 2x + 1$
14. $4x - 3 = 2x + 7$
15. $2x + 3 = 3x - 4$
16. $3x - 2 = x - 6$
17. $6x - 1 = 3x + 5$
18. $3x - 7 = x + 2$
19. $5x - 3 = x - 5$
20. $3x + 5 = 4x - 10$

Use the same methods as in Questions 1–20 to solve the following equations:

21. $3x - 2 = 2x - 1$
22. $4x - 1 = 2x + 5$
23. $-1 + 4x = 5 + 2x$
24. $1 - 3x = 2 - 4x$
25. $7 + x = 2x - 3$
26. $9x + 5 = 4x - 15$
27. $7x - 12 = 5x - 6$
28. $2x - 3 = 3x - 8$
29. $-3x + 2 = -5x + 8$
30. $2x + 5 = 4x - 4$

Exercise 12.12 *Use of indices*

Simplify each of the following:

1. $a \times a \times a \times a \times a$	**2.** $a \times a^4$	**3.** $a^2 \times a^3$
4. $b \times b \times b$	**5.** $b \times b^2$	**6.** $a \times a \times b$
7. $a^2 \times b$	**8.** $a \times ab$	**9.** $a \times a \times b \times b$
10. $a \times b \times a \times b$	**11.** $ab \times ab$	**12.** $a^2 \times ab$
13. $c^4 \times c^5$	**14.** $c^5 \times c^6$	**15.** $cd \times c^2$
16. $2 \times a \times a$	**17.** $3 \times 2 \times a$	**18.** $3 \times a \times 2 \times a$
19. $3a \times 2a$	**20.** $6a \times a$	**21.** $2a \times 3b$
22. $2a^2 \times ab$	**23.** $5a \times 2ab$	**24.** $3a^2 \times 4ab$

Simplify each of the following:

25. $\dfrac{a \times a \times a}{a \times a}$	**26.** $\dfrac{a \times a \times b}{a \times b}$	**27.** $\dfrac{a \times a \times b \times b}{a \times a \times b}$
28. $\dfrac{b \times a \times a \times b}{a \times b \times b}$	**29.** $\dfrac{a^2 \times b}{ab}$	**30.** $\dfrac{a^2b^2}{ab}$
31. $a^2b \div ab$	**32.** $a^4b \div a^2b$	**33.** $18b \div 6$
34. $18b \div b$	**35.** $18b \div 6b$	**36.** $7b \div 2b$
37. $x^2y \div xy$	**38.** $\dfrac{p^2}{pq}$	**39.** $6c \div \frac{1}{6}c$
40. $16ab^2 \div 8b$	**41.** $6a^8 \div 3a^5$	**42.** $\dfrac{6p \times 3pq}{9p}$
43. $\dfrac{8a \times 3b}{16b}$	**44.** $24abc \div 8ab^2$	**45.** $3a^4 \times 5ab \div a^5$
46. $3p^2q \div 6pq^2$	**47.** $\dfrac{6a + 7a^2}{3a}$	**48.** $\dfrac{4a^2b \times 2ab^2}{8b^3 \times a^2}$

Exercise 12.13 *Common factors and multiples*

Find the Highest Common Factor of each of the following pairs:

1. $3a,\ 3b$	**2.** $3a,\ ab$	**3.** $2a,\ 6$
4. $3ab,\ 6bc$	**5.** $a^2b,\ ac$	**6.** $5a,\ 10a^2b$
7. $5ab^2,\ 10b^2c$	**8.** $12cd^2,\ 4c^2d$	**9.** $x^5,\ x^3$
10. $6x^4,\ 2x$	**11.** $2x^2y,\ 4a^2b$	**12.** $9x^2y,\ 15xy^2$
13. $2\pi r,\ \pi r^2$	**14.** $\frac{1}{2}ah,\ \frac{1}{2}bh$	**15.** $\pi R^2,\ \pi r^2$
16. $6a^2,\ -2a^3$	**17.** $8ab,\ -4a$	**18.** $-10x^4y,\ -2x^3$

Find the Lowest Common Multiple of each of the following pairs:

19. $a,\ ab$	**20.** $3a,\ ab$	**21.** $a,\ a^5$
22. $a^2,\ a^4b$	**23.** $x^2y^5,\ x^3y^2$	**24.** $9x^4y^3,\ 3xy^4$
25. $21x^5,\ 7x^4y$	**26.** $3abc,\ 4cba$	**27.** $4l^2m^3,\ 12lm^2$
28. $a^4m^2,\ 5am^3$	**29.** $3a^4b^2,\ 12a^2b^3$	**30.** $a^4b^2,\ 3a^3b$
31. $2ax^2,\ 3ax$	**32.** $3x^2y^2,\ 9x^3y$	**33.** $4x^2y^3,\ 8xy^4$
34. $5m^2n^3,\ 10n^2p$	**35.** $9x^4y^5,\ 12x^3y^4$	**36.** $12cd^2e^4,\ 4de^2f$

By taking out the Highest Common Factor, insert brackets where possible in the following:

37. $2a+4$ **38.** $2n-4$ **39.** $2c+2d$ **40.** $2c+3c$
41. $2.6\times 3+2.6\times 7$ **42.** $ac+ad$ **43.** $bx-dx$ **44.** $7x-14y$
45. $6r-5s$ **46.** $6z-3z$ **47.** $2r-4rs$ **48.** p^2-p
49. $2xy-yz$ **50.** a^2-ax **51.** $6a^2-2ay$ **52.** $12pq+6p^2$
53. $180n-360$ **54.** $\frac{1}{2}ah+\frac{1}{2}bh$ **55.** $\frac{1}{2}t^2-\frac{1}{4}t$ **56.** $-2t+4at$

Factorise each of the following:

57. a^2-4a **58.** $5a^2-a^{10}$ **59.** $3a^5x-3a^5y$
60. $5ax-a^5y$ **61.** $-a^2b+a^3$ **62.** $3a^2x+3a^2y+3a^2xy$
63. $p^2x-2px+3px^2$ **64.** $6n^6+4n^4+2n^2$ **65.** $2x^4+3x^3+4x^2$
66. $15tv-5v^2t+3t^2v$ **67.** $9xy^2+18xy-6x^2y^2$ **68.** $15a^2-5ax+3x^2$
69. $a(x)+ay$ **70.** $a(x+1)+ay$ **71.** $a(x+1)+2(x+1)$

Exercise 12.14 *Equations with brackets*

Solve the following equations:

1. $2(x+3)=10$ **2.** $2(2a+1)=6$ **3.** $3(b-2)=9$
4. $4(a-1)=20$ **5.** $3(2-a)=-2$ **6.** $6=3(a-1)$
7. $2(d-5)=1$ **8.** $6(x+2)=10x$ **9.** $4+2(x-1)=7$
10. $3-(x+1)=1$ **11.** $x+2(x-1)=4$ **12.** $3+2(x-5)=2$
13. $5a-2=3(a+4)$ **14.** $4(x-4)=3x-4$ **15.** $3(x-7)=x+9$
16. $4(2x-5)=5x+1$ **17.** $6(x+3)=58-4x$ **18.** $5(2x-1)=3(x+3)$
19. $43(x-1)=13(3x+1)$ **20.** $2x+(x-5)=1$ **21.** $4(2a-1)+a=3a+8$
22. $3(a+1)+2(a-2)=9$ **23.** $2(a-1)-(a-3)=4$ **24.** $4(a-1)-2(a+2)=12$
25. $6(a-2)+1=4(a+10)$ **26.** $2(5c-1)-3(2+c)=6$

Exercise 12.15 *Inequalities*

Write out the following sentences as mathematical statements:

1. z is equal to 3 **2.** 7 is greater than 3
3. f is greater than 5 **4.** 7 is less than 8
5. x is less than 10 **6.** 4 is less than y
7. 7 is greater than h **8.** x is less than or equal to 6
9. t is less than or equal to 4 **10.** y is greater than or equal to 2

Write out the following mathematical statements in words:

11. $7>3$ **12.** $1<4$ **13.** $7>x$
14. $x<7$ **15.** $x>1$ **16.** $x<4$
17. $x<0$ **18.** $y>0$ **19.** $-3<-1$
20. $x<-1$ **21.** $x>-6$ **22.** $x\geqslant 0$
23. $x\leqslant 0$ **24.** $x\leqslant -3$ **25.** $x\geqslant -1$

26. For each of questions 13–18 and 20–25, give three values which satisfy the mathematical statement, taking your values from the set of integers only.

13

Angle calculations

Exercise 13.1 *Numerical calculations*

Find the size of the angles marked by small letters.
(In Questions 1–6 ABC is a straight line.)

1.

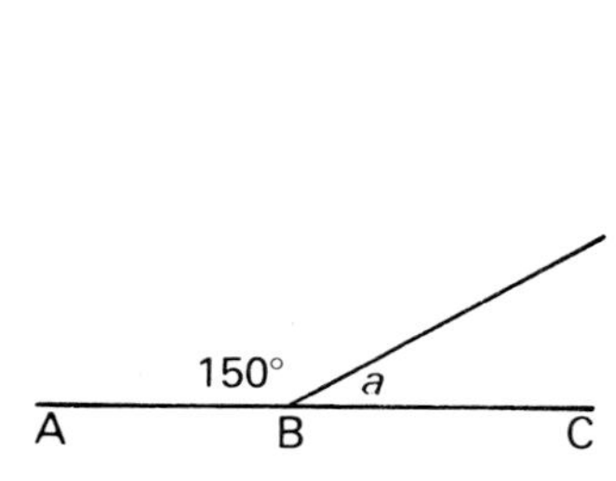

2.

A
B
b
65°
C

3.

C
c
43°
B
A

4.

5.

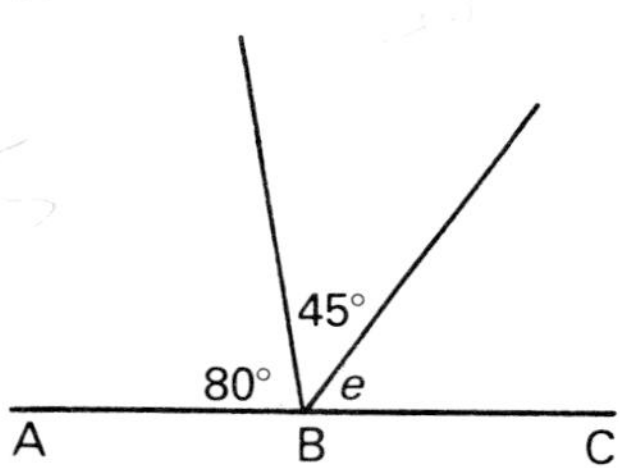

6.

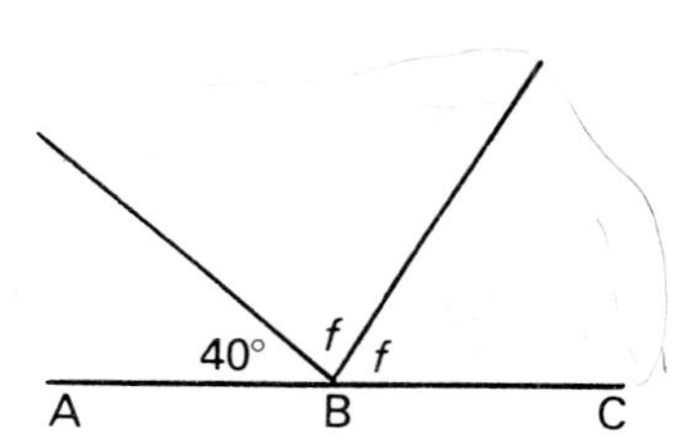

7.

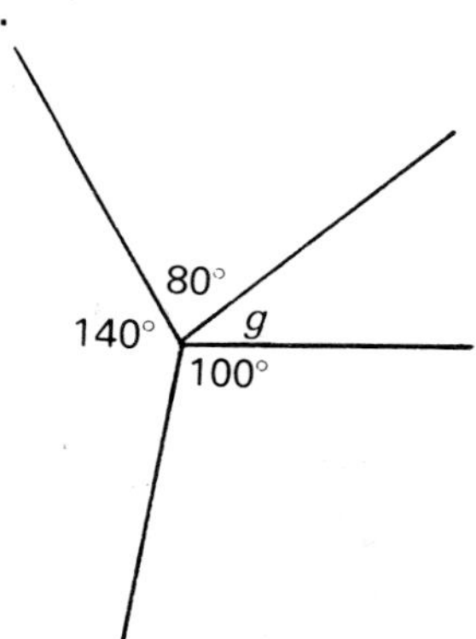

8.

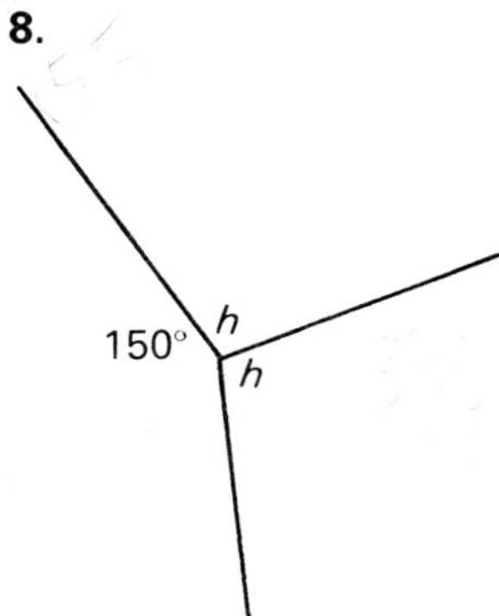

9.

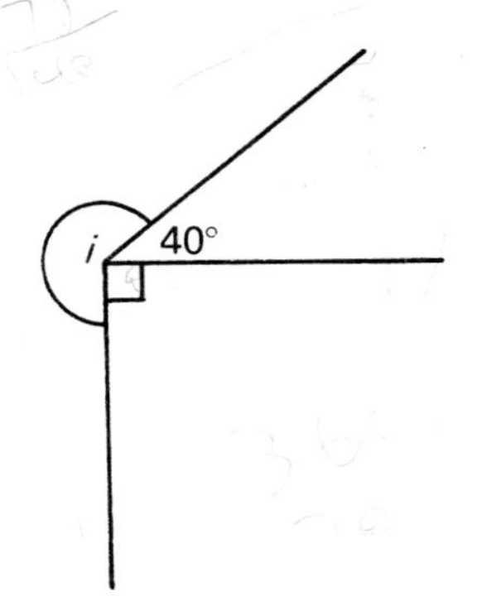

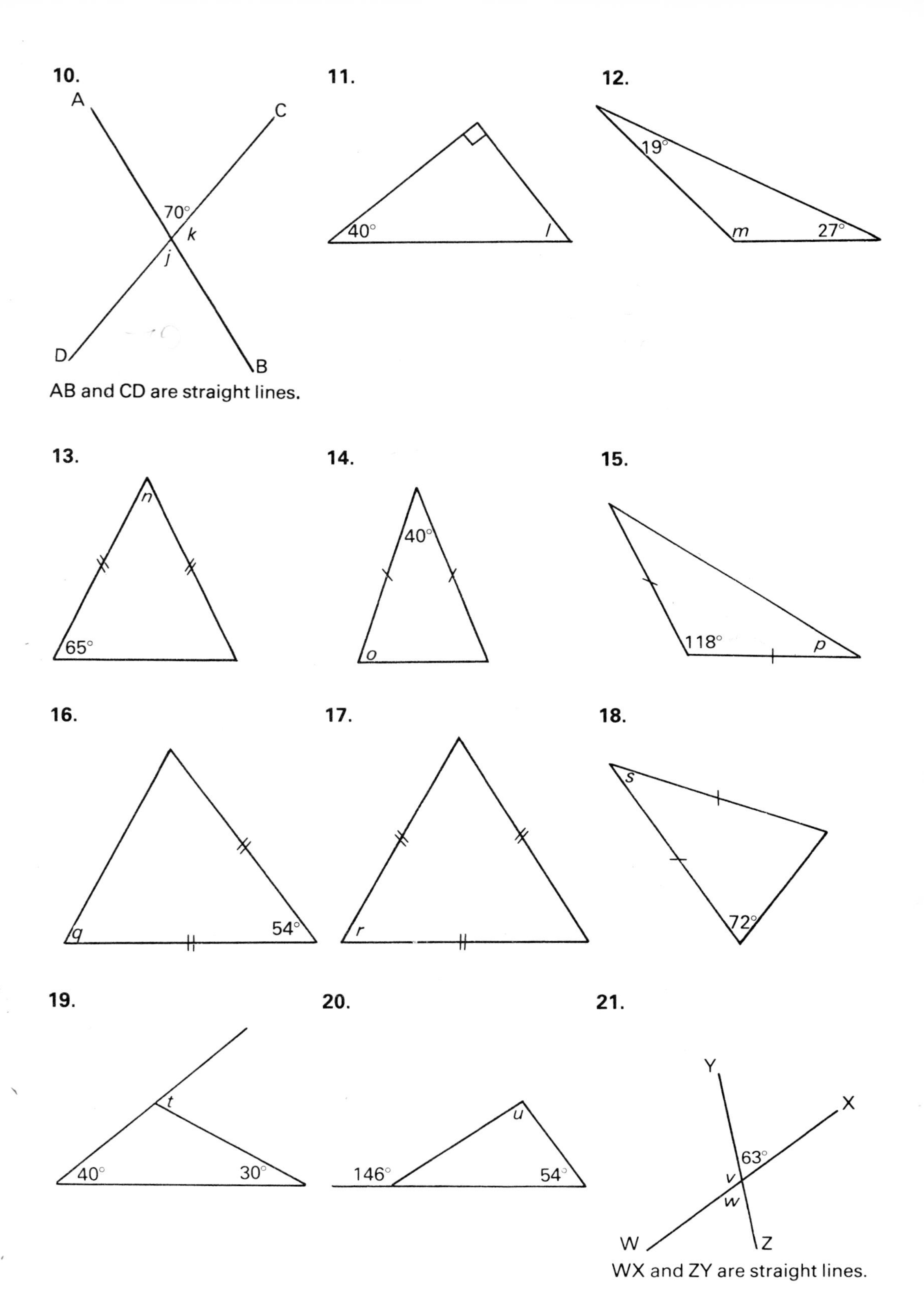
10.
A
C
70°
k
j
D
B
AB and CD are straight lines.
11.
40°
l
12.
19°
m
27°
13.
n
65°
14.
40°
o
15.
118°
p
16.
q
54°
17.
r
18.
s
72°
19.
t
40°
30°
20.
u
146°
54°
21.
Y
X
63°
v
w
W
Z
WX and ZY are straight lines.

22.

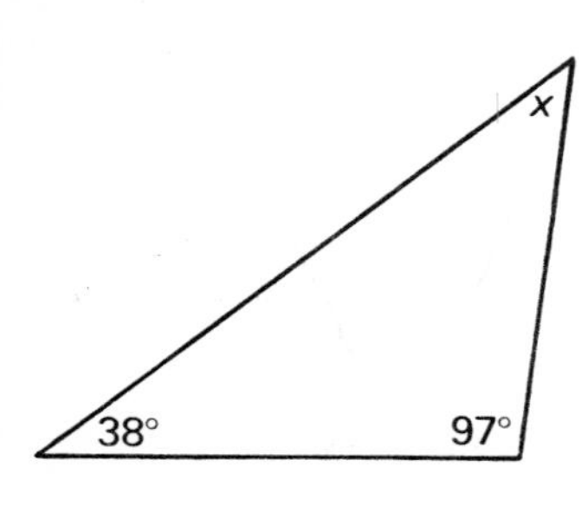

23.

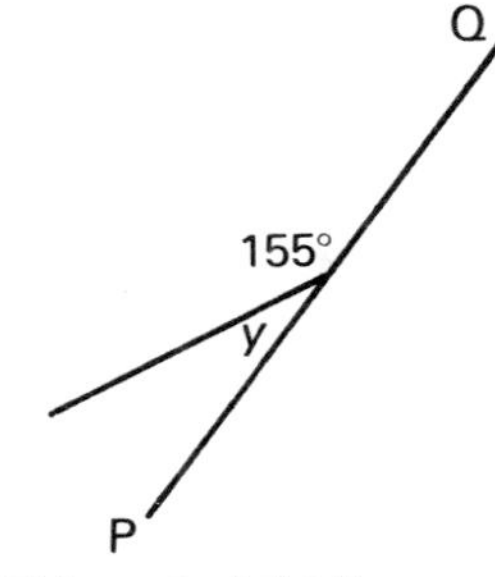

PQ is a straight line.

24.

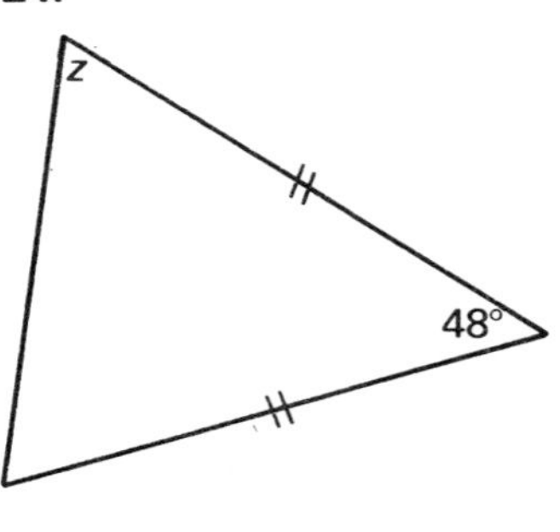

25.

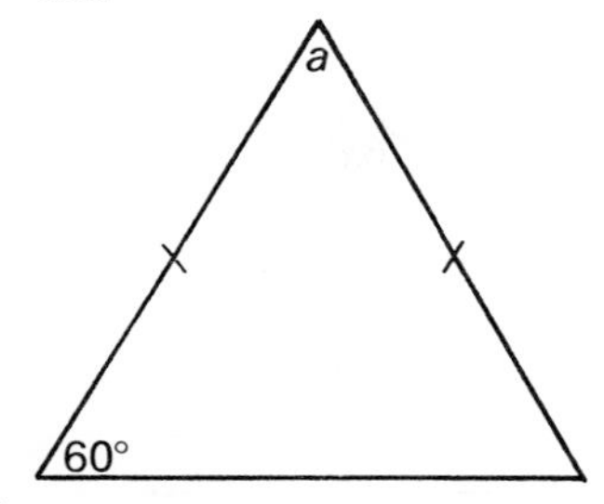

26.

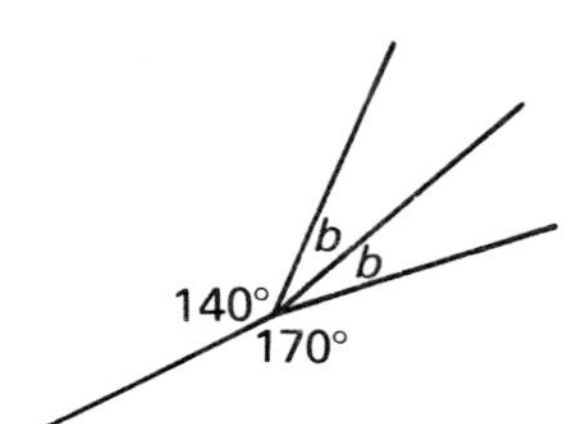

27.

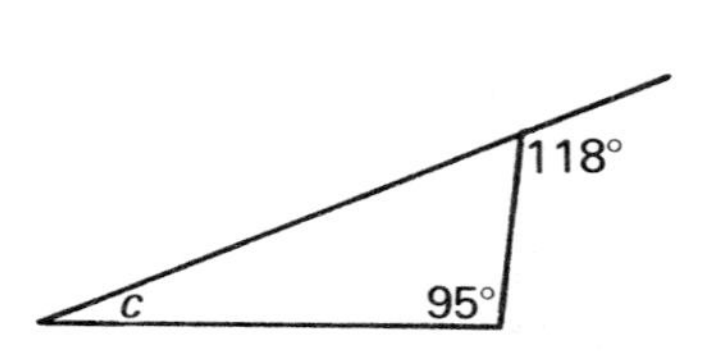

28.

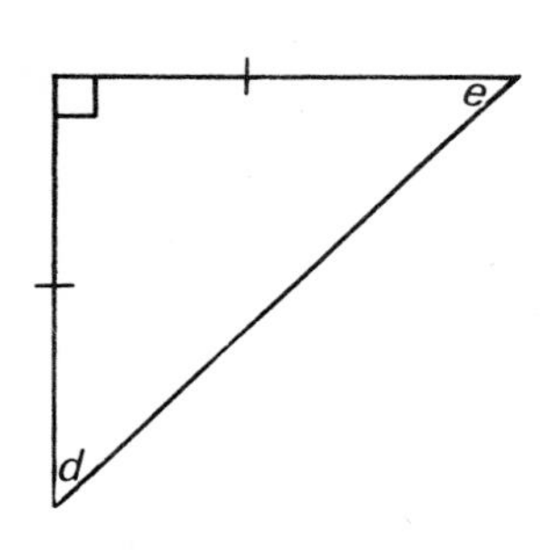

29.

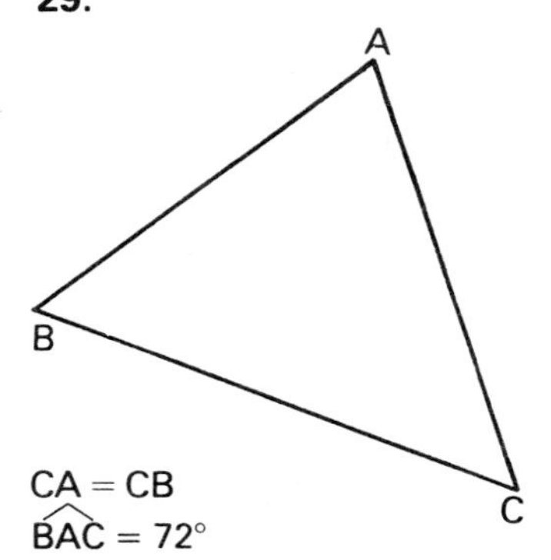

CA = CB
$\widehat{BAC} = 72°$

Calculate $\widehat{ACB}$ and reflex $\widehat{ABC}$.

30.

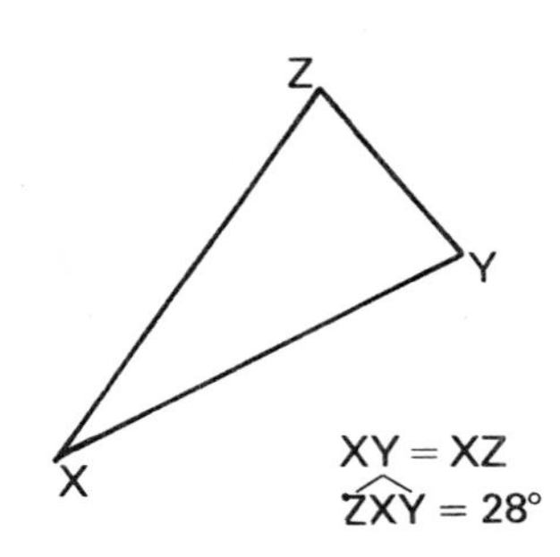

XY = XZ
$\widehat{ZXY} = 28°$

Calculate $\widehat{XYZ}$ and reflex $\widehat{XZY}$.

Exercise 13.2 *Algebraic calculations*

In each question find the value of x. Also write down which 'angle fact' you are using.

1.

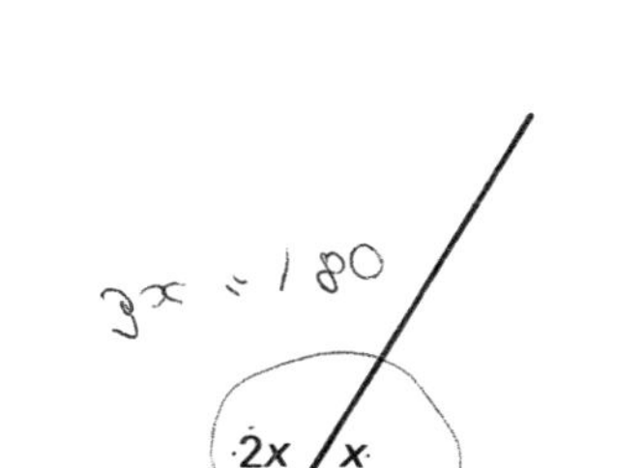

2.

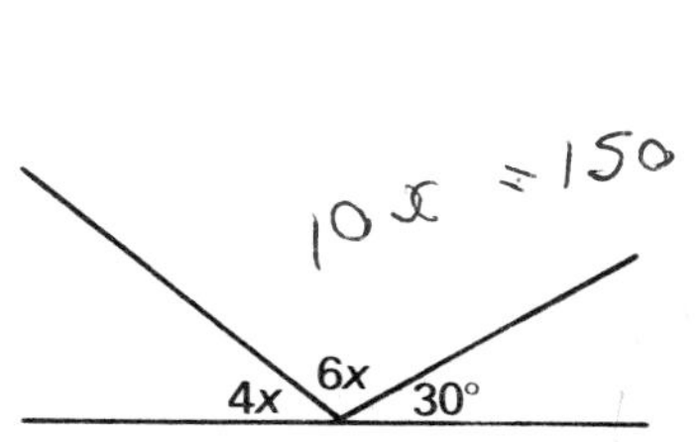

3.

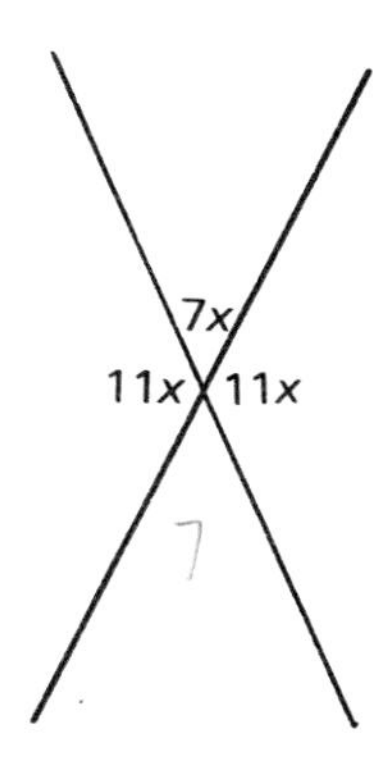

4.

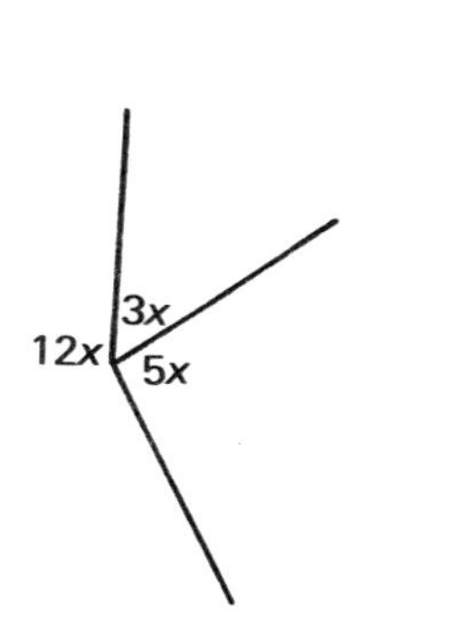

5.

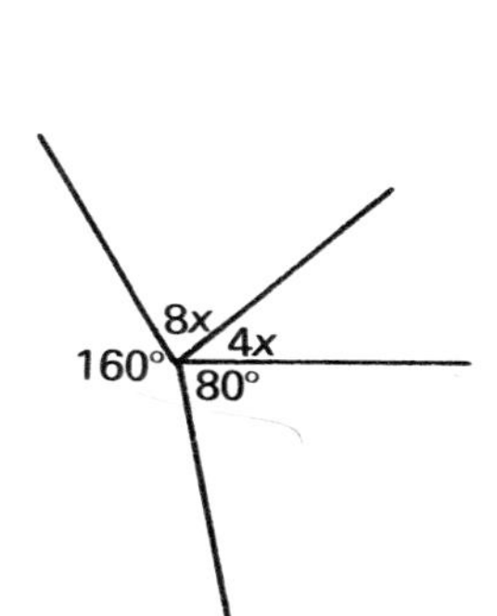

6.

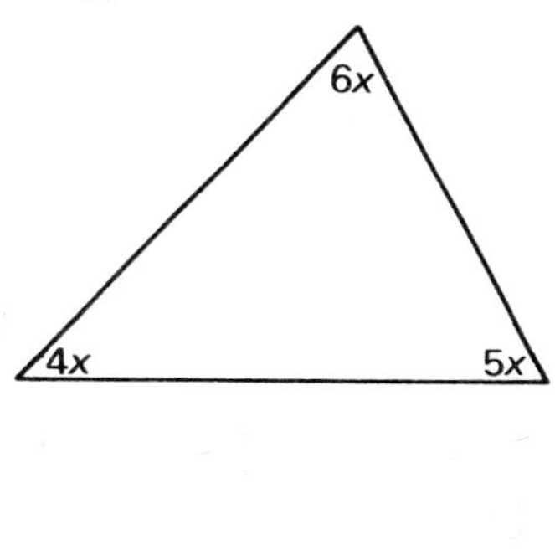

7.

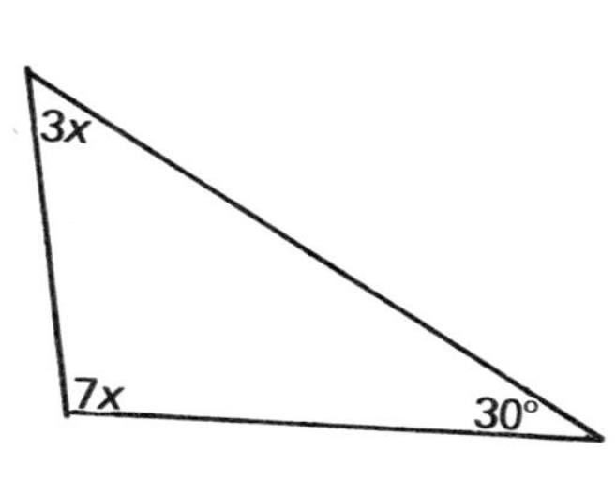

8.

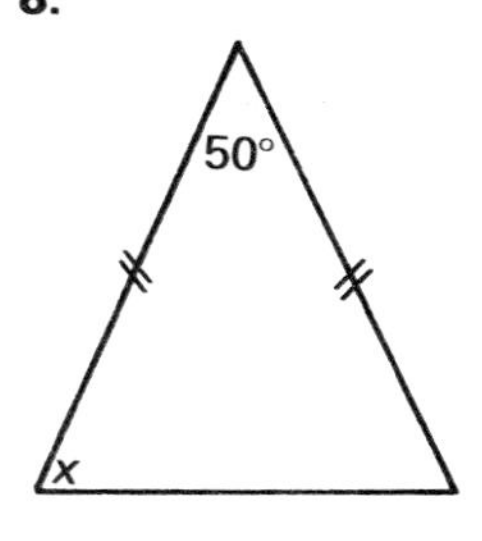

9.

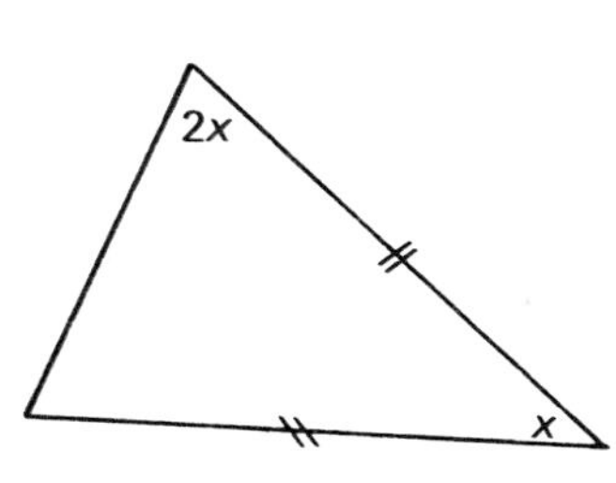

10.

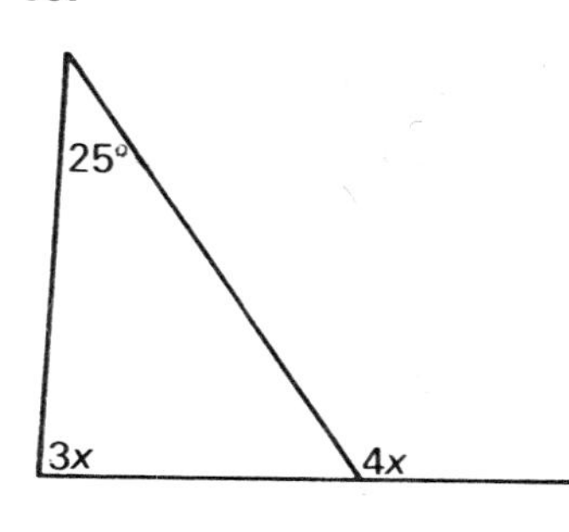

11.

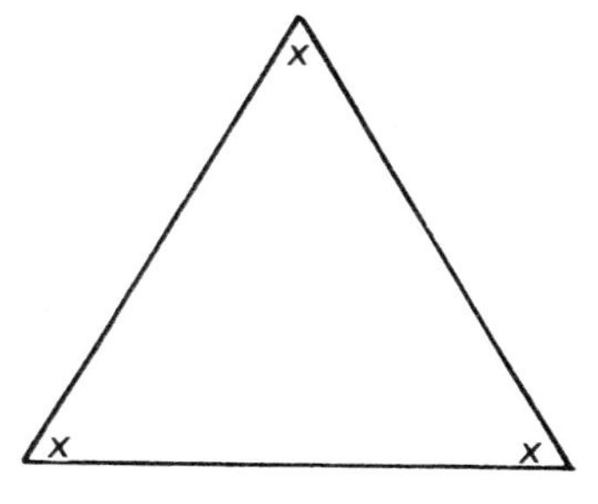

12.

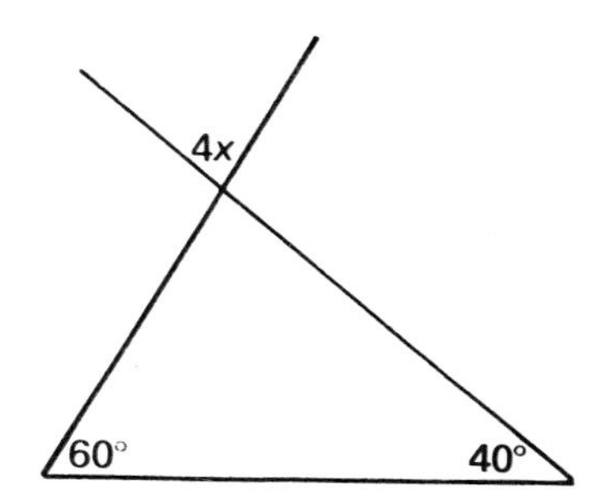

13.

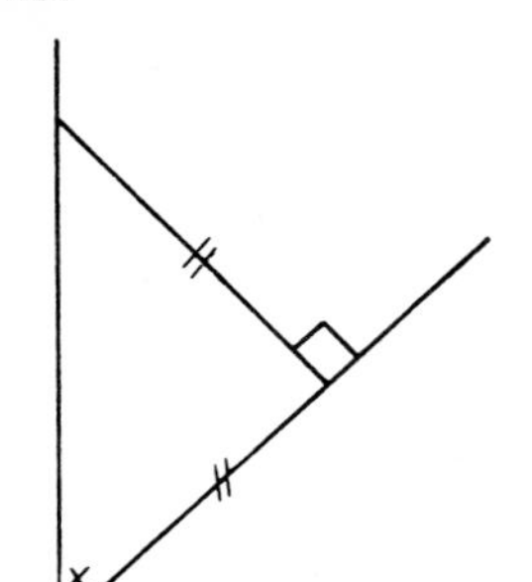

14.

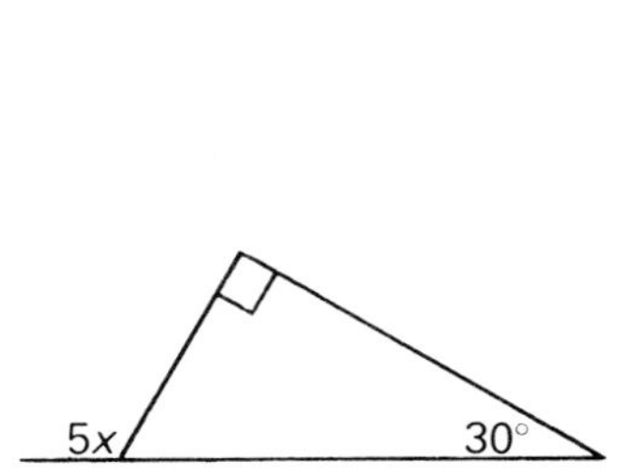

15.

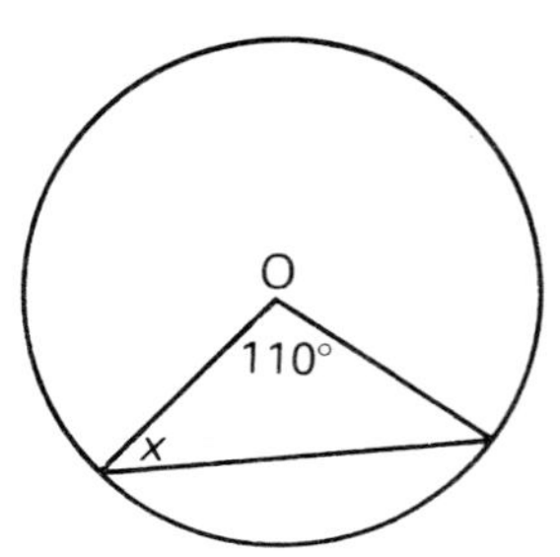

O is the centre of the circle.

Exercise 13.3 *Parallel lines*

Calculate the angles marked with small letters. Write down which 'angle fact' you are using.

1.

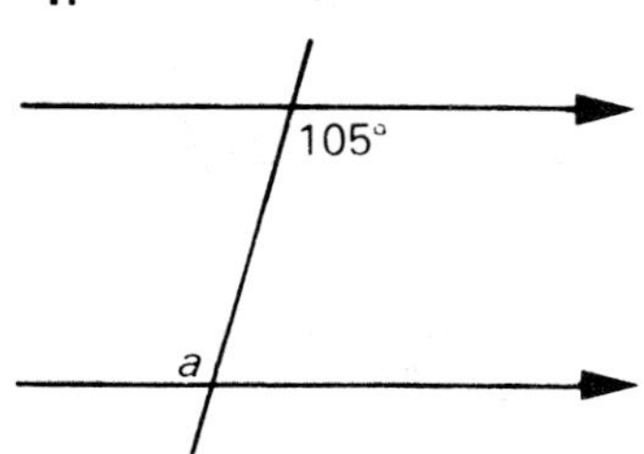

2.

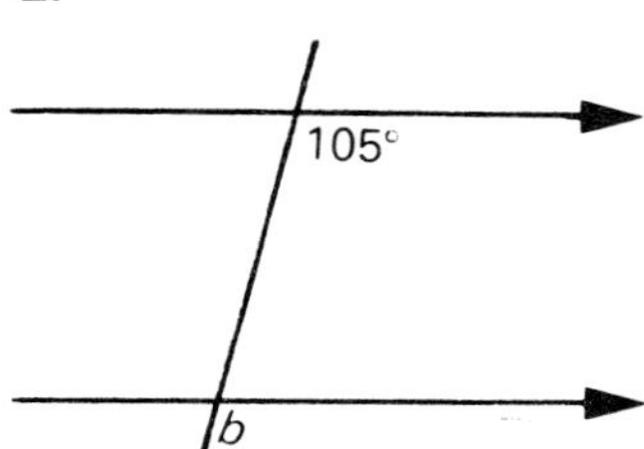

3.

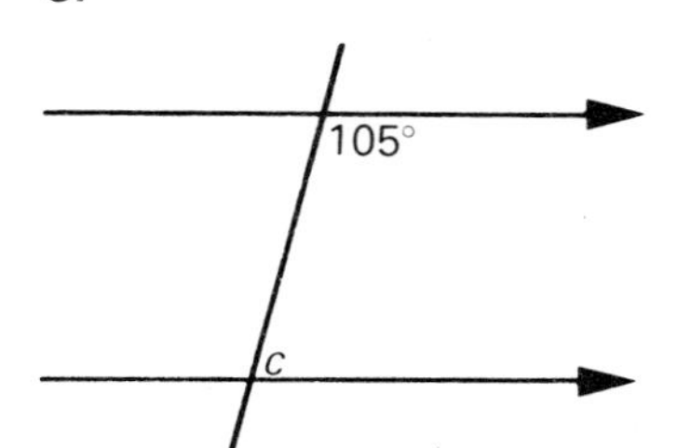

4.

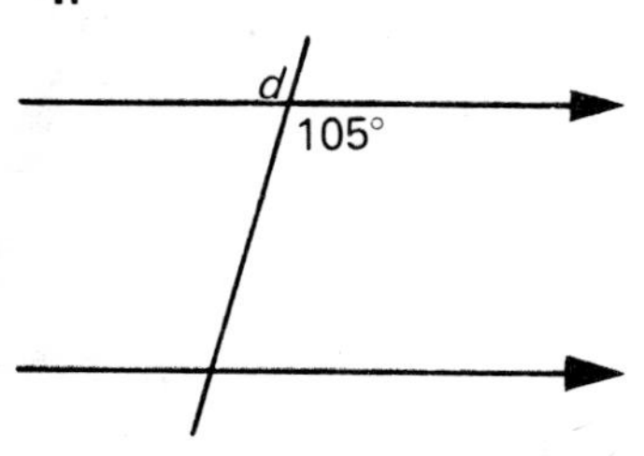

5.

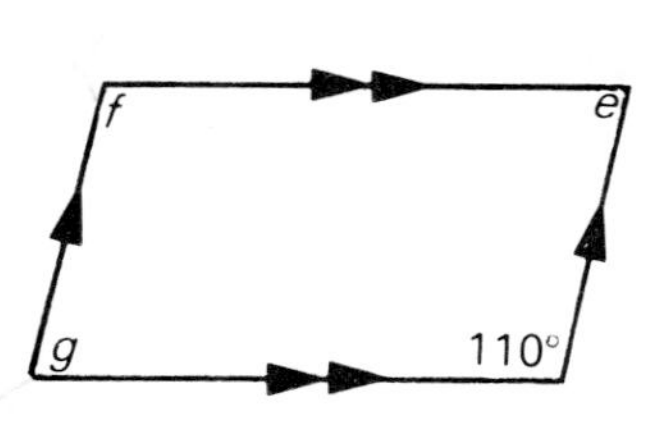

6.

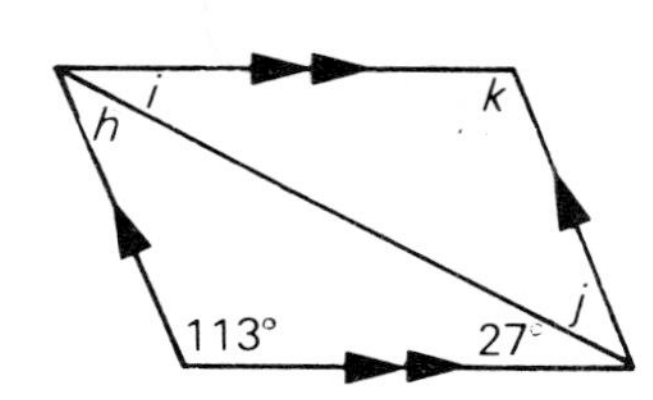

7.

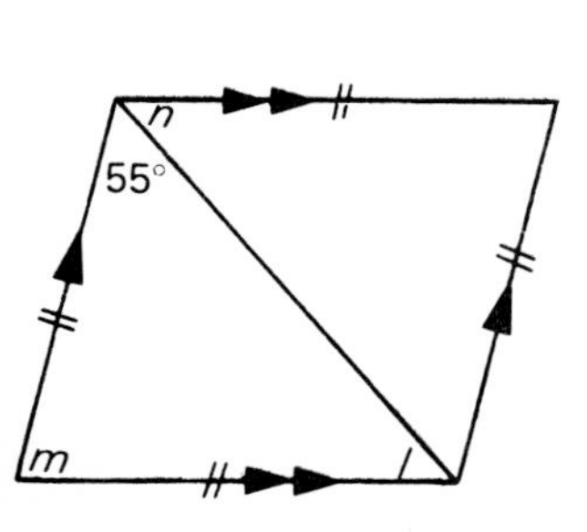

8.

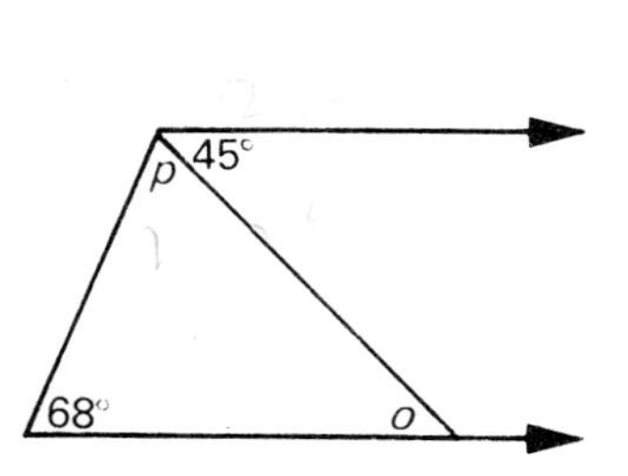

9.

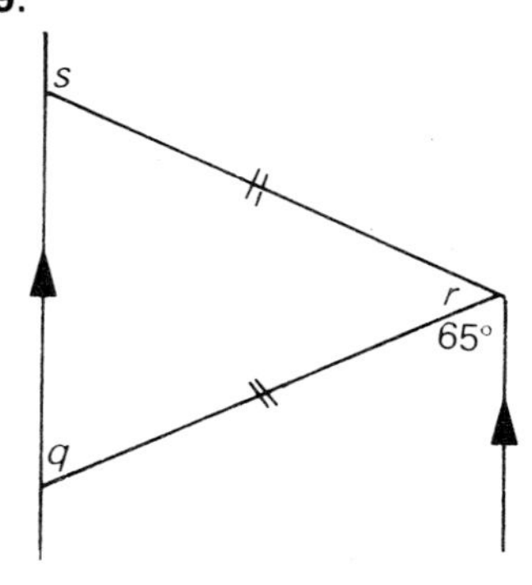

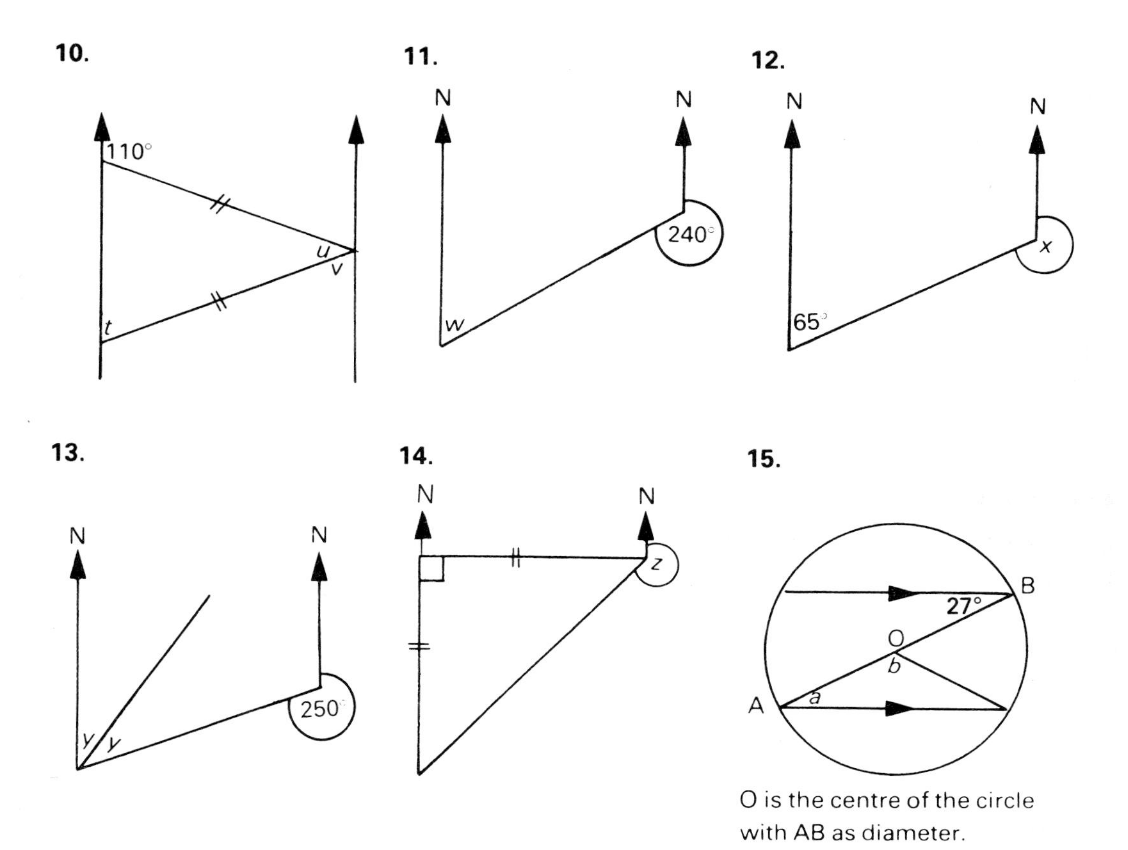

O is the centre of the circle with AB as diameter.

14

Construction of triangles

Exercise 14.1 *Lines*

Construct lines of the following length, lettering them as shown:

1. AB = 4 cm
2. CD = 11 cm
3. EF = 6.3 cm
4. GH = 9.5 cm
5. JK = 7.8 cm
6. LM = $8\frac{1}{2}$ cm
7. NO = 40 mm
8. PQ = 100 mm
9. RS = 53 mm
10. XY = 128 mm

Exercise 14.2 *Angles – circumflex notation*

For each example draw AB = 6 cm and construct the following angles:

1. $\hat{A} = 30°$
2. $\hat{A} = 150°$
3. $\hat{B} = 45°$
4. $\hat{B} = 135°$
5. $\hat{A} = 63°$
6. $\hat{B} = 107°$
7. $\hat{B} = 81°$
8. $\hat{A} = 85°$
9. $\hat{A} = 104°$
10. $\hat{B} = 95°$

Exercise 14.3 *Angles – three letter notation*

For each example draw XY = 5 cm and construct the following angles:

1. $\widehat{XYZ} = 73°$
2. $\widehat{YXZ} = 49°$
3. $\widehat{ZYX} = 64°$
4. $\widehat{XYZ} = 87°$
5. $\widehat{YXZ} = 115°$
6. $\widehat{YXZ} = 127°$
7. $\widehat{ZXY} = 17°$
8. $\widehat{XYZ} = 97°$
9. $\widehat{ZYX} = 79°$
10. $\widehat{YXZ} = 142°$

Exercise 14.4 *Naming angles*

Copy the figure. Mark the small letters given below in the correct angles:

$a = \widehat{BAD}$ $b = \widehat{ACB}$
$c = \widehat{ABD}$ $d = \widehat{CBE}$
$e = \widehat{BDC}$ $f = \widehat{FBA}$
$g = \widehat{FBE}$ $h = \widehat{BDA}$
$i = \widehat{CBD}$ $j = \widehat{ABC}$

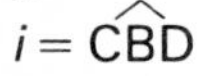

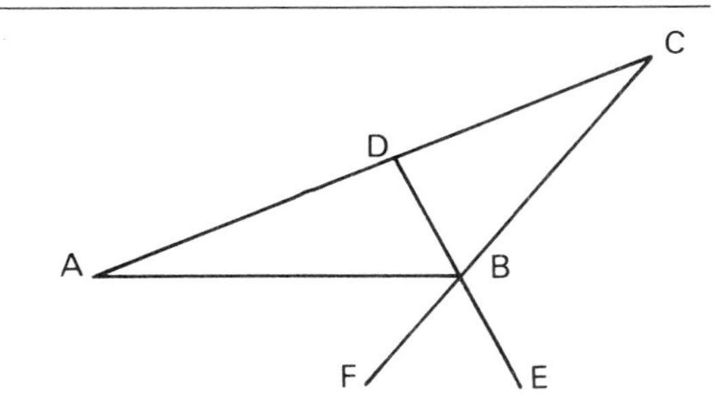

Exercise 14.5 *Construction of triangles*

Construct the following triangles using the given information, and give the measurements required:

1. $\triangle ABC$: $AB = 10$ cm, $AC = 7.4$ cm, $\hat{A} = 54°$. Measure (**a**) BC (**b**) $\hat{B}$.
2. $\triangle PQR$: $PQ = 7.2$ cm, $QR = 7.4$ cm, $\hat{Q} = 97°$. Measure (**a**) PR (**b**) $\hat{R}$.
3. $\triangle XYZ$: $XY = 8.7$ cm, $XZ = 8.1$ cm, $Y\hat{X}Z = 63°$. Measure (**a**) YZ (**b**) $X\hat{Y}Z$.
4. $\triangle ABC$: $AB = 6.3$ cm, $AC = 6.8$ cm, $B\hat{A}C = 119°$. Measure (**a**) BC (**b**) $A\hat{C}B$.
5. $\triangle RST$: $RS = 8.8$ cm, $ST = 6.6$ cm, $R\hat{S}T = 44°$. Measure (**a**) RT (**b**) $R\hat{T}S$.
6. $\triangle ABC$: $AB = 10$ cm, $\hat{A} = 49°$, $\hat{B} = 58°$. Measure (**a**) AC (**b**) $\hat{C}$.
7. $\triangle XYZ$: $XY = 7.8$ cm, $X\hat{Y}Z = 51°$, $Y\hat{X}Z = 59°$. Measure (**a**) YZ (**b**) $X\hat{Z}Y$.
8. $\triangle XYZ$: $XY = 6.4$ cm, $Y\hat{X}Z = 102°$, $X\hat{Y}Z = 43°$. Measure (**a**) YZ (**b**) $Y\hat{Z}X$.
9. $\triangle ABC$: $AB = 8$ cm, $A\hat{B}C = 47°$, $B\hat{A}C = 57°$. Measure (**a**) AC (**b**) $B\hat{C}A$.
10. $\triangle XYZ$: $XY = 8.5$ cm, $\hat{X} = 32°$, $\hat{Y} = 102°$. Measure (**a**) YZ (**b**) $\hat{Z}$.
11. $\triangle PQR$: $PQ = 10$ cm, $PR = 8$ cm, $QR = 6$ cm. Measure (**a**) $\hat{Q}$ (**b**) $\hat{R}$.
12. $\triangle ABC$: $AB = 9.8$ cm, $AC = 9.3$ cm, $BC = 5.3$ cm. Measure (**a**) $\hat{A}$ (**b**) $\hat{C}$.
13. $\triangle PQR$: $PQ = 8.2$ cm, $PR = 8.9$ cm, $QR = 6.7$ cm. Measure (**a**) $P\hat{Q}R$ (**b**) $P\hat{R}Q$.
14. $\triangle DEF$: $DE = 8.2$ cm, $EF = 11.7$ cm, $DF = 6.5$ cm. Measure (**a**) $E\hat{D}F$ (**b**) $D\hat{F}E$.
15. $\triangle ABC$: $AB = 8.4$ cm, $BC = 7.2$ cm, $AC = 11.4$ cm. Measure (**a**) $A\hat{B}C$ (**b**) $C\hat{A}B$.
16. $\triangle ABC$: $AB = 10$ cm, $B\hat{A}C = 53°$, $A\hat{B}C = 46°$. Measure (**a**) AC (**b**) $A\hat{C}B$.
17. $\triangle XYZ$: $XY = 9.8$ cm, $YZ = 7.8$ cm, $X\hat{Y}Z = 102°$. Measure (**a**) XZ (**b**) $X\hat{Z}Y$.
18. $\triangle PQR$: $PQ = 6.4$ cm, $PR = QR = 9.2$ cm. Measure (**a**) $P\hat{Q}R$ (**b**) $P\hat{R}Q$.
19. $\triangle ABC$: $AB = 11$ cm, $BC = 6.6$ cm, $A\hat{B}C = 108°$. Measure (**a**) AC (**b**) $B\hat{A}C$.
20. $\triangle XYZ$: $XY = 10.5$ cm, $X\hat{Y}Z = 55°$, $Y\hat{X}Z = 47°$. Measure (**a**) XZ (**b**) YZ.
21. $\triangle LMN$: $LM = 8.7$ cm, $\hat{L} = 60°$, $\hat{N} = 90°$ (Be careful). Measure the two unknown sides.

22. Construct $AB = 7.5$ cm, $A\hat{B}C = 110°$, $B\hat{A}C = 100°$. Can you draw a triangle with this information? If not, explain why.
23. Can you construct $\triangle XYZ$ if you know that $XY = 10$ cm, $XZ = 5$ cm and $YZ = 3$ cm? If not, explain why.
24. Construct $\triangle ABC$, given that $AB = 9.5$ cm, $BC = 5.5$ cm and $B\hat{A}C = 30°$. Is it possible to construct more than one triangle with this information?

Exercise 14.6 *Mediators*

Draw 3 lines of different length and each time construct their mediators (perpendicular bisectors), using ruler and compasses only. Check your construction by measurement.

Exercise 14.7 *Angle bisectors*

Draw the following angles: 40°, 75°, 108°, 140°, 200°. In each case construct the bisector of the angle using ruler and compasses only. Check your construction by measurement.

Exercise 14.8 *Perpendiculars*

Copy the figures below (use tracing paper). In each case construct, using ruler and compasses only, the line from P which is perpendicular to AB.

1. (**a**)

(**b**)

2. Copy the figure below (use tracing paper).

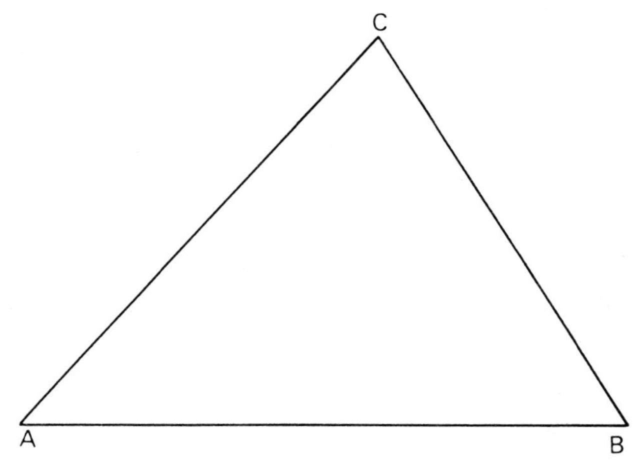

Using ruler and compasses only:

(**a**) Construct the perpendicular from B which cuts AC at D. Measure BD.

(**b**) Construct the line from A which is perpendicular to BC. Let it cut BC at E; measure BE.

3. Copy the figure below (use tracing paper).

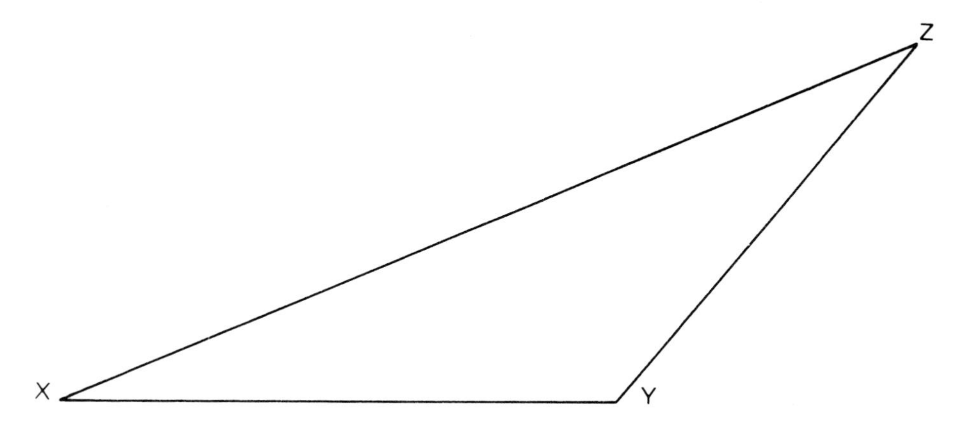

Using ruler and compasses only:

(**a**) Construct the perpendicular from Y which cuts XZ at O. Measure YO.

(**b**) Construct the perpendicular from Z which cuts XY produced at P. Measure YP.

4. PQ is 10 cm. R is a point on PQ such that PR = 4 cm. Erect a line at R which is perpendicular to PQ.

Exercise 14.9 *Triangles – including ruler and compass constructions*

(All mediators, angle bisectors and perpendiculars are to be constructed using ruler and compasses only.)

1. (**a**) Construct $\triangle ABC$ when $AB = 10.5$ cm, $AC = 7.3$ cm and $B\widehat{A}C = 40°$.

(**b**) Construct the bisector of angle B and the line from C which is perpendicular to AB.

(**c**) Let these two lines meet at O. Measure CO.

2. (**a**) Construct $\triangle DEF$ when $DE = 7.8$ cm, $E\widehat{D}F = 125°$ and $D\widehat{E}F = 25°$.

(**b**) Construct the mediator of EF and the bisector of $D\widehat{F}E$.

(**c**) Let these two lines meet at P. Measure EP.

3. (**a**) Construct the triangle in which $JK = 8.3$ cm and $JL = KL = 6.3$ cm.

(**b**) Construct the mediators of JK and JL.

(**c**) Label the point where the mediators meet as O.

(**d**) With centre O and radius OK draw a circle.

(**e**) Measure the radius of the circle.

4. (**a**) In triangle MNO, $MN = 12.5$ cm, $NO = 8.6$ cm and $\widehat{N} = 51°$. Construct the triangle.

(**b**) Construct the bisectors of angles M and N.

(**c**) Mark as P the point where the bisectors meet.

(**d**) Construct the perpendicular from P which cuts MN at Q.

(**e**) With centre P and radius PQ draw a circle.

(**f**) Measure the radius of the circle.

5. (**a**) Construct $\triangle PQR$ when $PQ = 10.7$ cm, $QR = 7.2$ cm and $P\widehat{Q}R = 114°$.

(**b**) Construct the line through Q which is perpendicular to PR.

(c) Construct the bisector of $\widehat{PRQ}$.
(d) Let these two lines meet at O.
(e) Join PO.
(f) Measure $\widehat{POR}$.

6. (a) Construct $\triangle ABC$ when $AB = BC = AC = 8$ cm.
(b) Construct the mediator of AB.
(c) Construct the bisector of angle B.
(d) Construct the perpendicular from A to BC.
(e) What do you notice about the mediator, the angle bisector and the perpendicular?

7. (a) Construct $\triangle TUV$ when $TU = 7$ cm, $UV = 8$ cm and $TV = 12.5$ cm.
(b) Construct the perpendicular from V which cuts TU produced at W.
(c) Construct the perpendicular bisector of TV which meets VW produced at X.
(d) Label the mid point of TV as M.
(e) Construct the bisector of angle T which cuts MX and VX at Y and Z respectively.
(f) Measure YZ, XZ, UW and UM.

8. Using ruler and compasses only, construct $\triangle ABC$ when $AB = 8$ cm, $\widehat{BAC} = 90°$ and $\widehat{ABC} = 60°$. Measure AC.

15

Polygon facts

Exercise 15.1 *Facts of simple polygons*

1. What is meant by a polygon?
2. Give the names of polygons with these numbers of sides:
   (a) 3 (b) 4 (c) 5
   (d) 6 (e) 8 (f) 10
3. What is meant by a regular polygon?
4. What special name is given to a three-sided regular polygon?
5. What name is given to a regular four-sided polygon?
6. What do the interior angles of a three-sided polygon add up to?
7. What do the interior angles of a four-sided polygon add up to?
8. Give simple sketches to show the difference between the interior and exterior angles of a polygon.
9. What connection is there between the number of sides and the number of lines of symmetry of a regular polygon?

Exercise 15.2 *Triangle facts*

1. What is meant by a scalene triangle?
2. Draw neat sketches to show what is meant by each of these:
   (a) an acute-angled triangle
   (b) an obtuse-angled triangle
   (c) a right-angled triangle
   (d) an isosceles triangle
   (e) an equilateral triangle.
3. How many obtuse angles does an obtuse-angled triangle have?
4. How many acute angles does an acute-angled triangle have?
5. What is the size of each angle in an equilateral triangle?
6. Draw neat sketches to show each of these:
   (a) an acute-angled isosceles triangle
   (b) an obtuse-angled isosceles triangle
   (c) a right-angled isosceles triangle.
7. How many facts are needed if a scalene triangle is to be drawn exactly? What has this to do with congruence?
8. List the different sets of facts needed if a scalene triangle is to be drawn exactly. Does this include three angles? If not, why not?

Exercise 15.3 *Quadrilateral facts*

1. Give a neat sketch to show what is meant by a convex quadrilateral. Explain your drawing.

2. Describe as briefly as possible, with the aid of a sketch, what is meant by each of these:
   (**a**) a parallelogram
   (**b**) a rhombus
   (**c**) a rectangle
   (**d**) a square
   (**e**) a kite
   (**f**) an arrow-head
   (**g**) a trapezium.

   Mark on your sketches any lines that are equal or parallel as well as any lines of symmetry that the figures may have.
3. What name is given to a quadrilateral with two pairs of parallel sides?
4. What name is given to a quadrilateral with two pairs of parallel sides and all its sides equal?
5. What name is given to a quadrilateral with two pairs of parallel sides and all its angles equal?
6. What name is given to a four-sided figure with two pairs of parallel sides and all its sides and angles equal?
7. What name is given to a four-sided figure with only two pairs of adjacent sides equal?
8. What name is given to a four-sided figure with a pair of parallel sides?
9. Draw a sketch to show what is meant by an isosceles trapezium.

16

Coordinates

Exercise 16.1 *Coordinates – first quadrant*

1. Give the coordinates of the lettered points in the following diagrams:

(**a**)

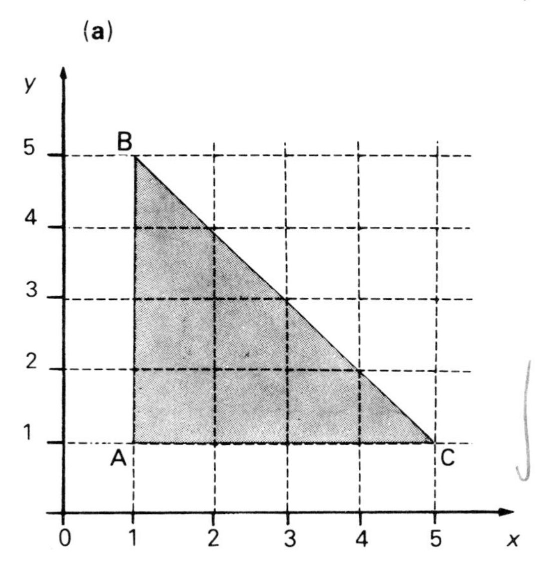

(**b**)

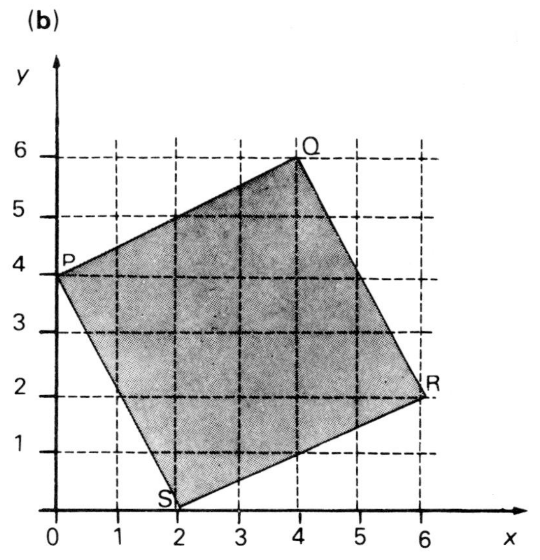

(**c**)

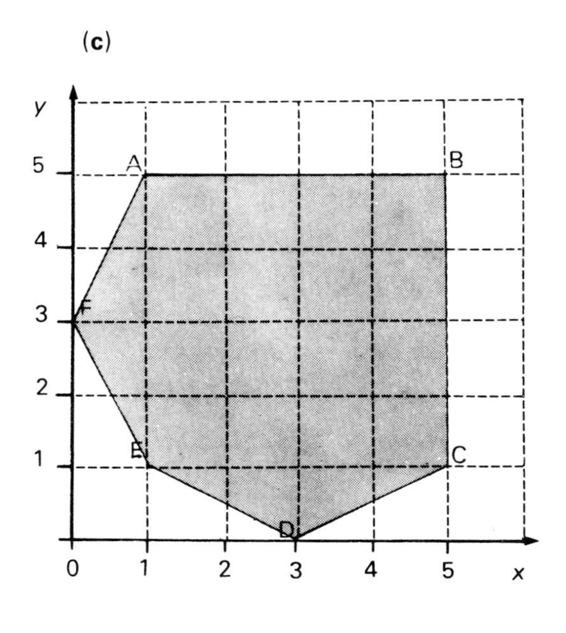

(**d**)

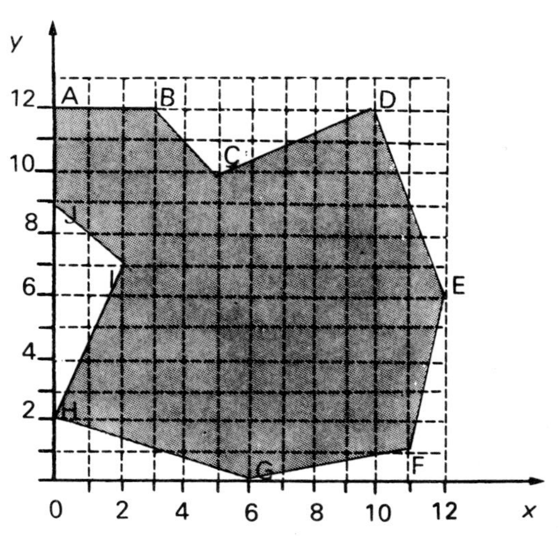

(e)

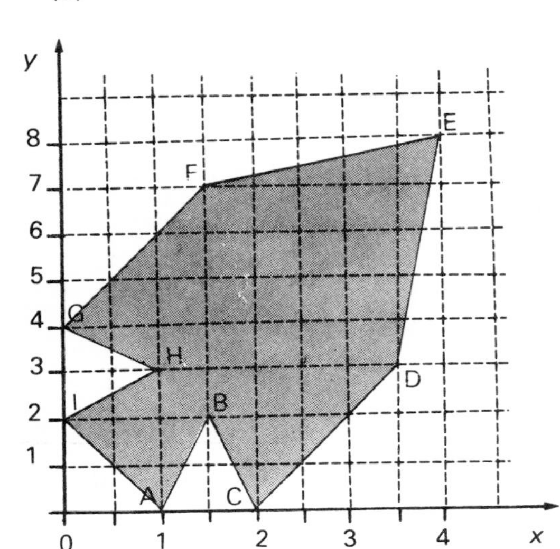

(f)

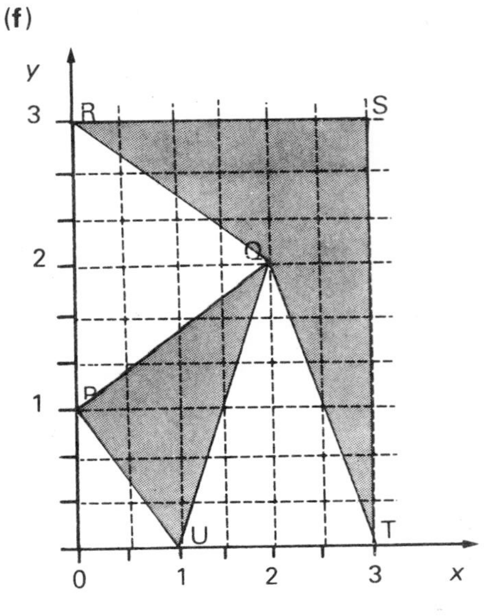

2. (**a**) What are the coordinates of the points A, B, C, D, E, F, G, H, I, J in the figure opposite?
   (**b**) Join these in order, finally joining J to A.
   (**c**) Dot in any lines of symmetry the figure may have.
   (**d**) Give the coordinates of two points other than E and J that lie on the line or lines of symmetry. What do you notice about their coordinates?

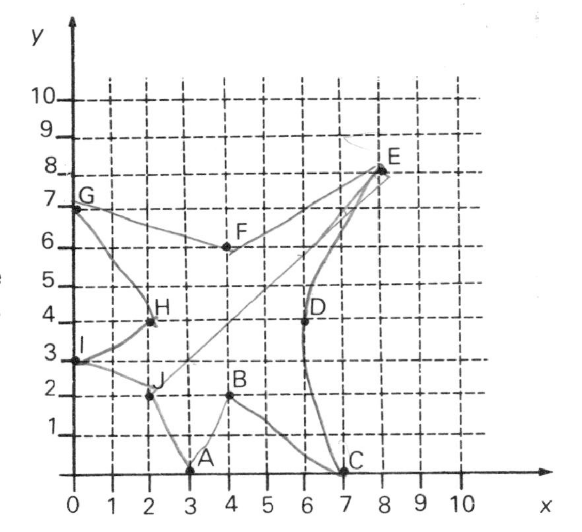

3. (**a**) What are the coordinates of the points A, B, C, D, E, F, G, H, I, J, K, L, M, N, O, P?
   (**b**) Join the points in that order and then finally join P to A.
   (**c**) Dot in any lines of symmetry the figure may have.
   (**d**) How many lines of symmetry has the figure?
   (**e**) How does this compare to the line symmetry of the square?

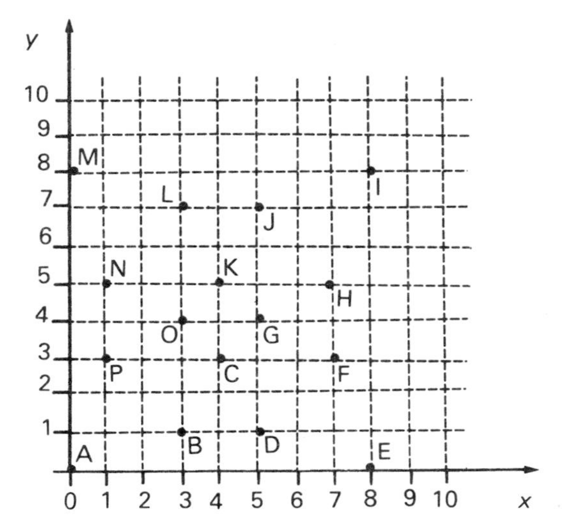

4. **(a)** A and P are two of the vertices of a square AXPY. What are the coordinates of the other two vertices?

(b) A and P are two of the vertices of a square APXY. What are the coordinates of the other two vertices? (Careful here!)

(c) What is the area of AXPY in square units?

(d) What is the area of APXY in square units?

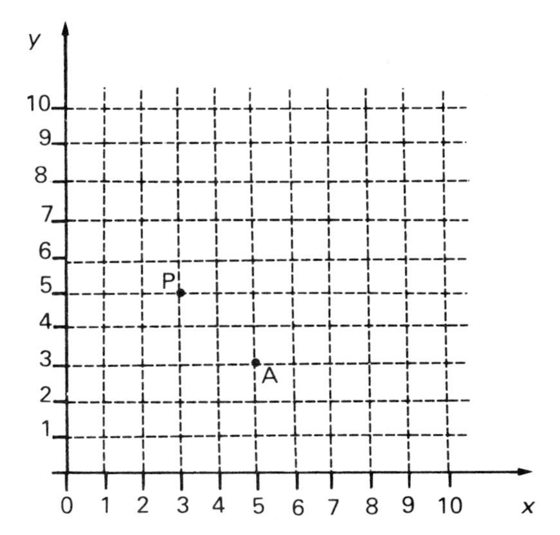

5. **(a)** What are the coordinates of the points A, B, C, D, E, F, G, H, I, J, K, L?

(b) Join the points in order and finally join L to A.

(c) Dot in any lines of symmetry that the figure may have.

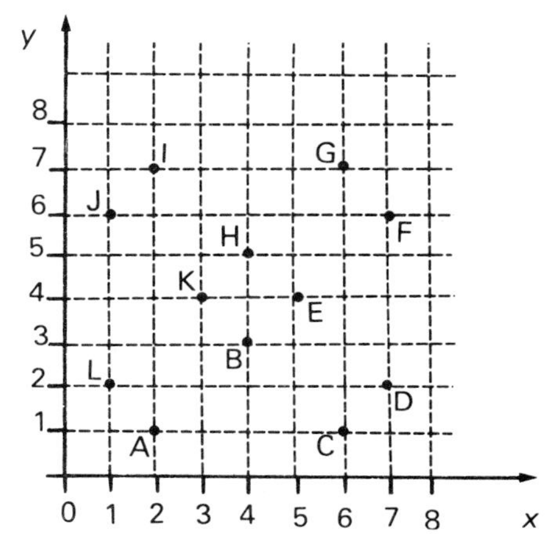

6. **(a)** Plot the points A(4,0), B(0,4), C(4,8), D(8,4) and then join them in order, finally joining D to A. What shape do you get?

(b) If you now rotate the figure about the point R(4,4) through an angle of 90° clockwise, which points would the points A, B, C and D map on to? If you like you may use the 'mapping' notation,
i.e. A →, B →, C →, D →.

(c) How many angles of 90° in the same direction would you have to turn altogether, about the point (4,4), until A → A, B → B, C → C, D → D?

7. **(a)** Plot the points A(1,1), B(8,1), C(6,0), D(2,0), E(4,9), F(4,1), G(9,2).

(b) Join them in the order A → B, B → C, C → D, D → A, A → E, E → F, F → G, G → E.

(c) Find the total area of the shape in square units.

(d) Is the triangle EFG isosceles? Give reasons for your answer.

8. **(a)** Plot the points A(2,0), B(3,3), C(0,3), D(2,6), E(2,8), F(3,10), G(5,9), H(5,7), I(10,7), J(6,4), K(6,0).

(b) What name do you give to the shape ABJ? Find its area in square units.

(c) What name do you give to the shape BJHD? Find its area.

(d) What name do you give to the shape DHGE? Find its area.

(e) What name do you give to the shape BCHI? Can you find its area? If so, show how.

(f) Is the triangle BCD equilateral, isosceles, or neither? Give reasons for your answer.

9. (a) Plot the points P(0,5), Q(8,0), R(8,10). Join them to form triangle PQR.

(b) Mark the mid-points of PQ, QR and PR and join them to form the triangle XYZ.

(c) What is the ratio of the area of triangle XYZ to that of triangle PQR?

(d) Find the mid-points of the sides XY, YZ and XZ. Letter them A, B, C respectively. Join them to form the triangle ABC.

(e) What is the ratio of the area of triangle ABC to that of triangle PQR?

10. Plot the points O(0,0), A(0,8), B(8,8), C(8,0).

(a) Join the points together in that order, finally joining C to O.

(b) Mark the mid-points of the lines OA, AB, BC, and CO, labelling them P, Q, R, S, respectively. Join them to form the square PQRS. How does the area of this square compare to that of the original square?

(c) If you then repeat the operation by letting the mid-points of the sides PQ, QR, RS and SP be W, X, Y and Z respectively, how does the area of WXYZ compare to that of OABC?

Exercise 16.2 *Coordinates – four quadrants*

1. For each of the figures below, give the coordinates of the lettered points:

(a)

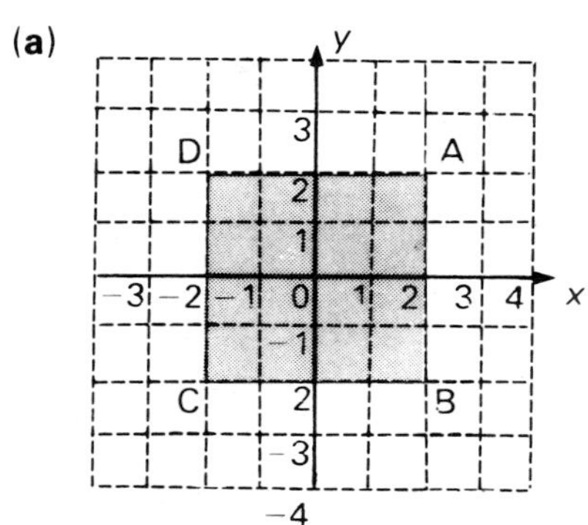

(b)

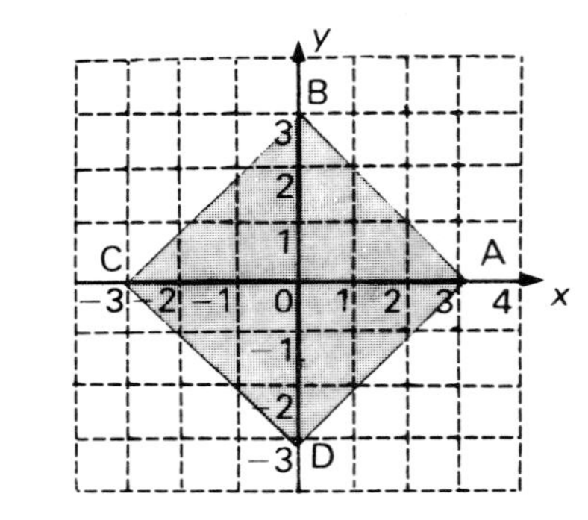

(c)

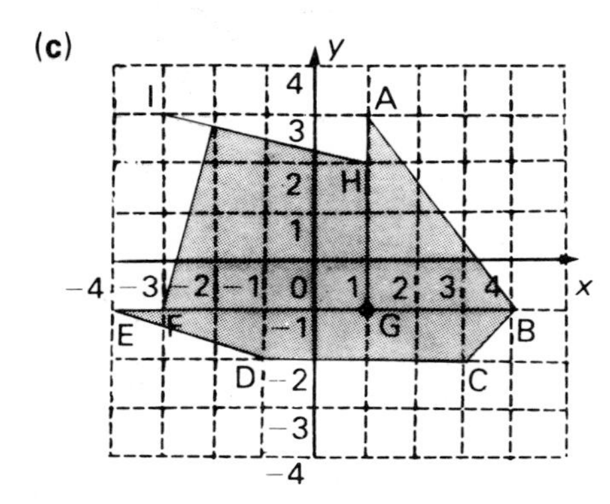

(d)

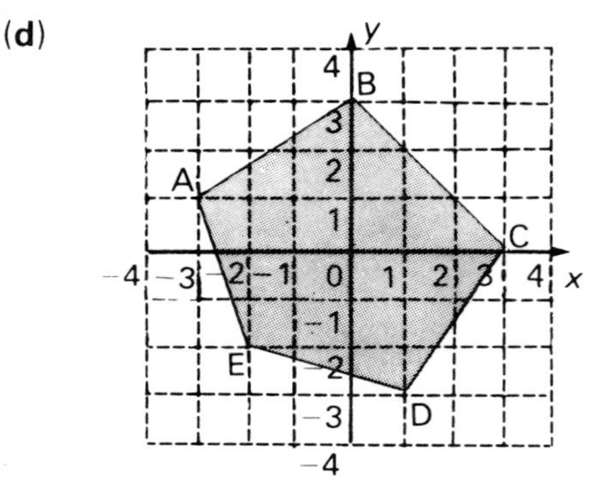

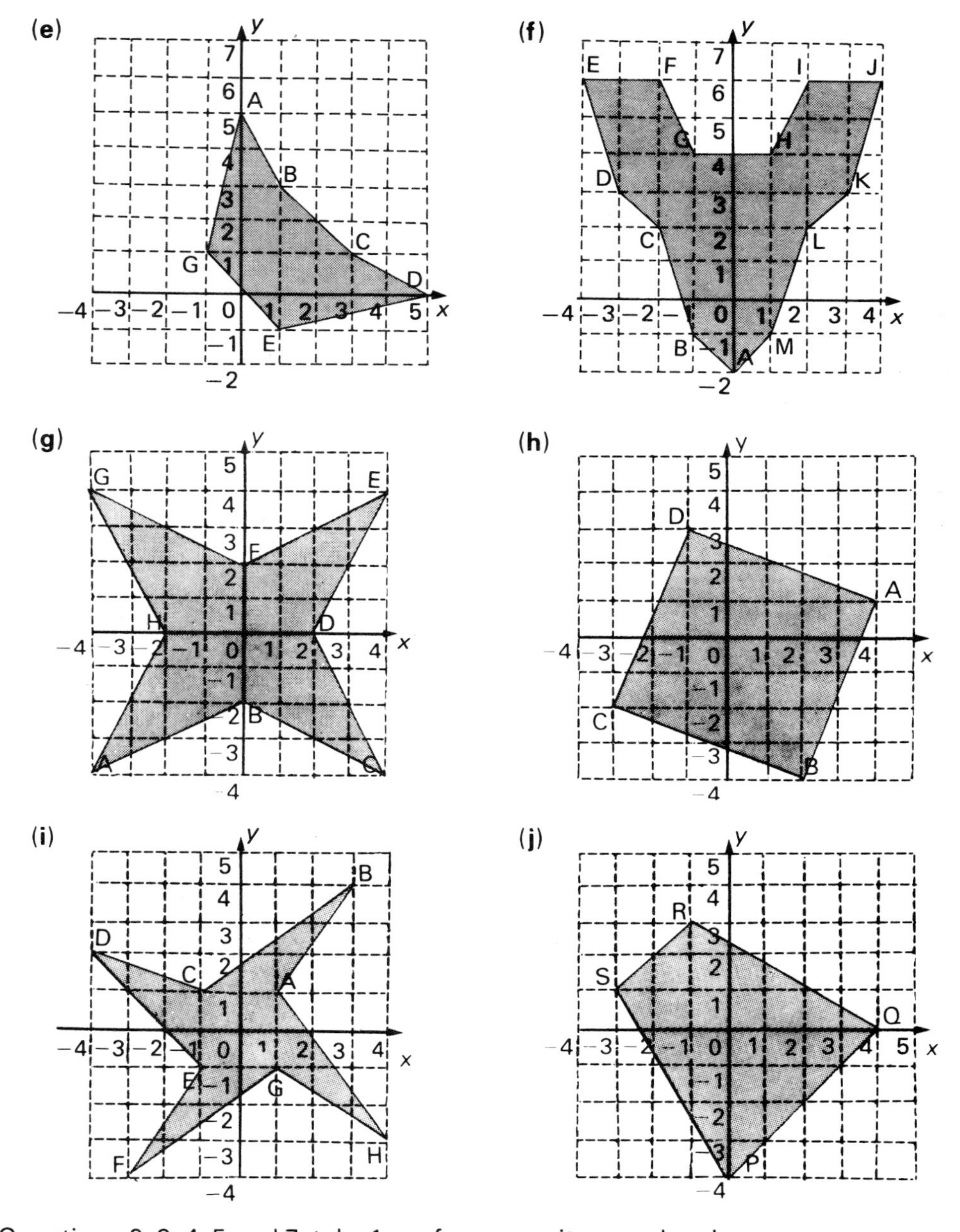

In Questions 2, 3, 4, 5 and 7, take 1 cm for one unit on each axis.

2. You will need to number both axes from -5 to $+5$.

(**a**) Plot the points A(1,1), B(1,5), C(5,1) to form the triangle ABC.

(**b**) Plot the points P(-1,-1), Q(-5,-1), R(-1,-5) to form the triangle PQR.

(**c**) Dot in any lines of symmetry that the figure as a whole may have.

(**d**) Has the figure as a whole any rotational symmetry?

3. You will need to number both axes from -5 to $+5$.

(**a**) Plot the points P(2,2), Q(1,-2), R(-2,-1), S(-1,3).

(**b**) Join the points in order to form the quadrilateral PQRS.

(**c**) What is true about these pairs of lines: **(i)** PS and QR **(ii)** SR and PQ?

(**d**) Do the coordinates of P, S and Q, R help you decide about your answer to (c)?

(**e**) What name is given to this kind of quadrilateral?

4. You will need to number both axes from -4 to $+4$.

(**a**) Plot the points A(2,4), B(4,2), C(2,0), D(4,−2), E(2,−4), F(0,−2), G(−2,−4), H(−4,−2), I(−2,0), J(−4,2), K(−2,4), L(0,2).

(**b**) Join them in that order, finally joining L to A.

(**c**) Are the axes lines of symmetry of the figure?

(**d**) Dot in any other lines of symmetry that you think the figure may have.

(**e**) What is the area of the shape in cm^2? (Does the symmetry of the figure help here?)

5. The axes should be numbered from -3 to $+3$ in this question.

(**a**) Plot the points P, Q, R and S as in question 3.

(**b**) Also plot the points A(−2,3), B(2,3), C(2,−2) and D(−2,−2) and join them to form the rectangle ABCD.

(**c**) What is the area of ABCD in cm^2?

(**d**) What fraction of the area of ABCD is the area of PQRS?

6. (**a**) What are the coordinates of the points A, B, C, D, E, F, G?

(**b**) Join the points in order and then join G to A.

(**c**) Dot in any lines of symmetry that the figure may have.

(**d**) By drawing a rectangle round the figure, find the area of the shape ABCDEFG.

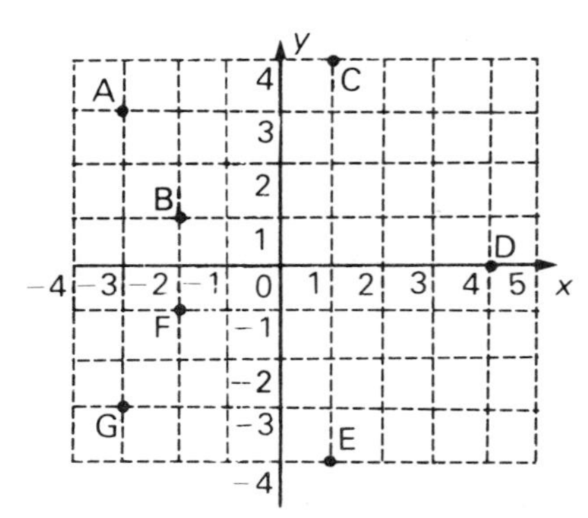

7. The axes should be numbered from -4 to $+4$ in this question.

(**a**) Plot the points A(4,0), B(−4,4), C(−4,−4) to form the triangle ABC.

(**b**) Plot the points X(−4,0), Y(4,−4), Z(4,4) to form the triangle XYZ.

(**c**) What is the area of triangle XYZ in cm^2?

(**d**) Is triangle ABC equilateral or isosceles? Give reasons for your answer.

8. From the figure on the right:

(**a**) Give the coordinates of the points A, B, C, D, E, F, G, H and join them in that order, finally joining H to A.

(**b**) Are the axes lines of symmetry?

(**c**) Dot in any other lines that you think are lines of symmetry.

(**d**) What is the area of BDFH in square units?

(**e**) What is the area of triangle BCD in square units?

(**f**) Use the symmetry of the figure to find the area of the figure ABCDEFGH in square units.

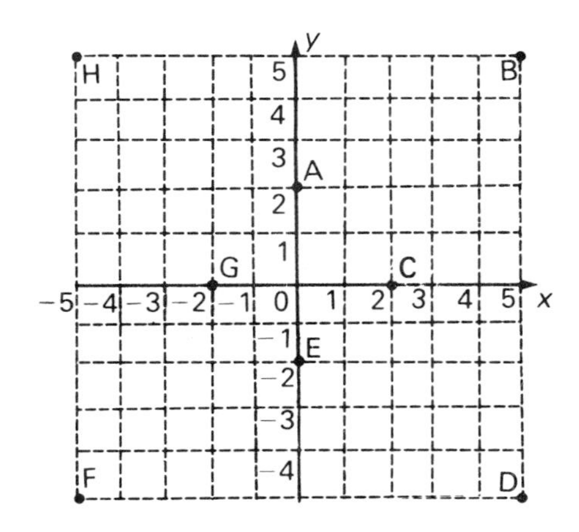

9. (a) What name do we give to the shape in Question 1(b)?
   (b) How many lines of symmetry has the figure in Question 1(a)?
   (c) What is the area of triangle AGB in Question 1(c)?
   (d) About which line is the figure in Question 1(f) symmetrical?
   (e) What is the order of rotational symmetry of the figure in Question 1(b)?
10. (a) What name do we give to the shape BCDE in Question 1(c)?
   (b) How many lines of symmetry has the figure in Question 1(g)?
   (c) What is the area of BCDE in Question 1(c)?
   (d) What name do we give to the shape in Question 1(j)?
   (e) What is the area of the shape in Question 1(h)?
   (You might find it helpful to enclose it with a square.)

Exercise 16.3 *Lines parallel to the axes and to* $y = \pm x$

1. Plot the points A(2,−1), B(2,5), C(2,0), D(2,−4).
   Draw a line through these points.
   (a) Give the coordinates of two other points on this line.
   (b) What do you notice about the x-coordinate in each case?
   (c) The line through these points is the line $x = ?$
   (d) If a point is to lie on this line does it matter what the y value is?
   (e) If you could extend the graph far enough would the point (2,100) lie on this line?
   (f) Does the point (3,4) lie on this line? Give reasons for your answer.
   (g) Can you suggest another line upon which the point B(2,5) lies?
2. Plot the points P(−1,5), Q(−1,0), R(−1,−1), S(−1,−4).
   (a) Give the coordinates of two other points on this line.
   (b) What do you notice about the x-coordinate in each case?
   (c) The line through these points is the line $x = ?$
   (d) If a point is to lie on this line, does it matter what the value of y is?
   (e) Will the point (−2,4) lie on this line? Give a reason for your answer.
   (f) Can you suggest at least one line that the point (−1,−1) will lie on, other than that in (c)?
3. Plot the points W(−1,3), X(−4,3), Y(1,3), Z(4,3).
   Draw a line through these points.
   (a) Give the coordinates of two other points on this line.
   (b) What do you notice about the y-coordinate in each case?
   (c) The line through all these points is the line $y = ?$
   (d) Would the point (100,3) lie on this line? Give reasons for your answer.
   (e) Upon which other line do you think that the point X(−4,3) would lie?
4. Plot the points K(−4,−2), L(0,−2), M(1,−2), N(4,−2).
   Draw a line through the points.
   (a) Give the coordinates of two other points on the line.
   (b) What do you notice about the y-coordinate in each case?
   (c) The line that can be drawn through these points is the line $y = ?$
   (d) Will the point (1,1) lie on this line? Give reasons for your answer.
   (e) Is this line parallel to the x axis or the y axis?
5. (a) Give two lines parallel to the axes upon which the point (3,1) will lie.
   (b) Give two lines parallel to the axes upon which the point (−2,4) will lie.
   (c) Give three lines upon which the point (3,3) will lie.
   (d) Give three lines upon which the point (0,0) will lie.

(**e**) Give three lines upon which the point $(-3,3)$ will lie.

(**f**) Upon which two lines parallel to the axes will the point (a,b) lie?

6. (**a**) Give the coordinates of three points lying on the line $x=2$.

(**b**) Give three points lying on the line $y=0$.

(**c**) Give the coordinates of three points lying on the line $x=y$.

(**d**) Give the coordinates of three points lying on the line $x=-y$.

(**e**) Can you give three points lying on the line $x-y=0$?

(**f**) Can you give three points lying on the line $x-y=4$?

(**g**) Can you give the coordinates of three points lying on the line $x+y=1$?

7. Draw axes with the x and y values going from -5 to $+5$.

(**a**) Plot the points A(4,4), B(4,−4), C(−4,−4), D(−4,4). Join them in order, finally joining D to A. Draw in the diagonals of the figure ABCD.

(**b**) Give the equation of the line AC.

(**c**) Give the equation of the line BD.

(**d**) The figure should have two other lines of symmetry. What are their equations?

(**e**) What are the equations of the sides AB, BC, CD and DA?

8. (**a**) Which axis runs parallel to the line $x=2$?

(**b**) Which axis runs parallel to the line $y=2$?

(**c**) Give two other lines running through the point (2,2).

17

Rotation and reflection

Exercise 17.1 *Line symmetry*

1. Take a rectangular piece of paper and fold it along the dotted line as shown in Fig. 1.
   Cut a pattern along the longest unfolded edge as shown in Fig. 2.
   Unfold the paper. You should have a design that is evenly balanced about the fold (Fig. 3). This pattern has line symmetry. The line of symmetry is the fold line.

Fig. 1

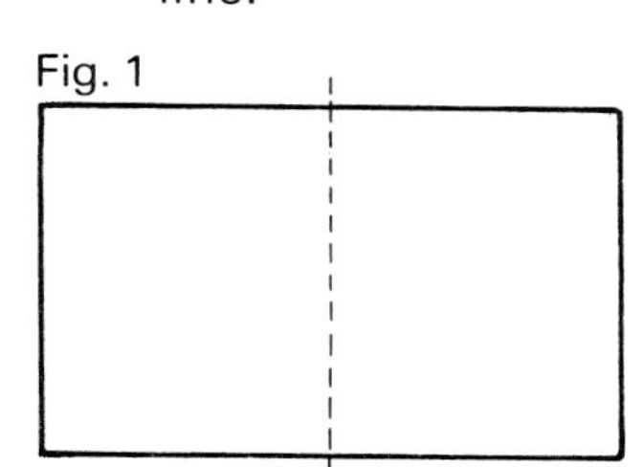

Fig. 2

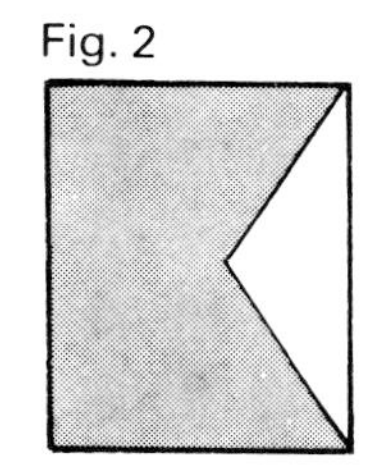

Fig. 3

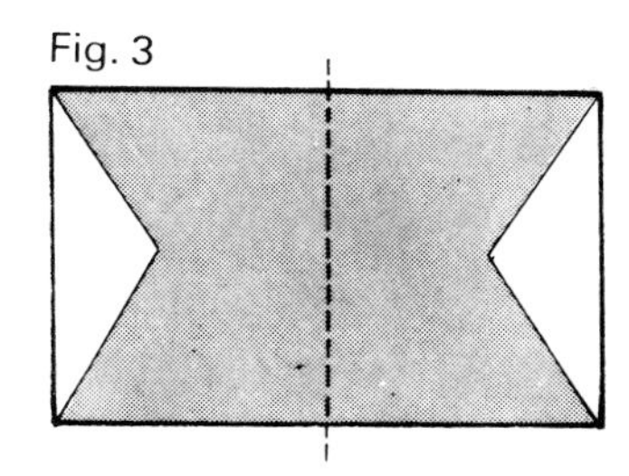

   Now repeat the idea, still with only one fold, with a more complicated design.
   Mark any point P on one side of the fold and mark the point it maps onto (touches) when you have folded the paper. Letter this (image) point P′.
   When you have unfolded the paper, join PP′ with a straight line. What do you notice about this line and the line of symmetry?
2. Once again take a rectangular piece of paper (do not use a square, as this might give a special result). Fold the paper along the dotted line AB as shown (Fig. 1).
   Then fold again along the line CD (Fig. 2).
   Cut a pattern along the unfolded edges (Fig. 3).
   Unfold the paper to give a pattern that is symmetrical about both the fold lines.
   Thus this pattern has two lines of symmetry.

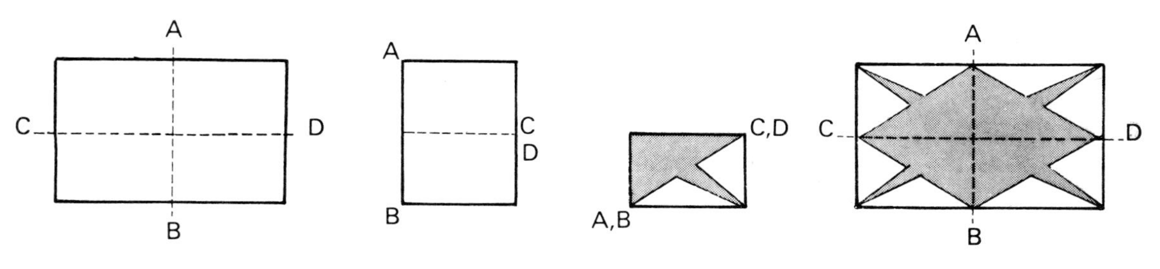

   Repeat the idea, using two folds, with a more complicated design.
   Mark any point P in one quarter of the paper. Mark the points it maps onto (touches) when you have folded the paper. Letter these (image) points P′, P″ and P‴.

When you have unfolded the paper, join all four points together with straight lines. What do you notice?

3. Copy the following shapes and dot in any lines of symmetry:

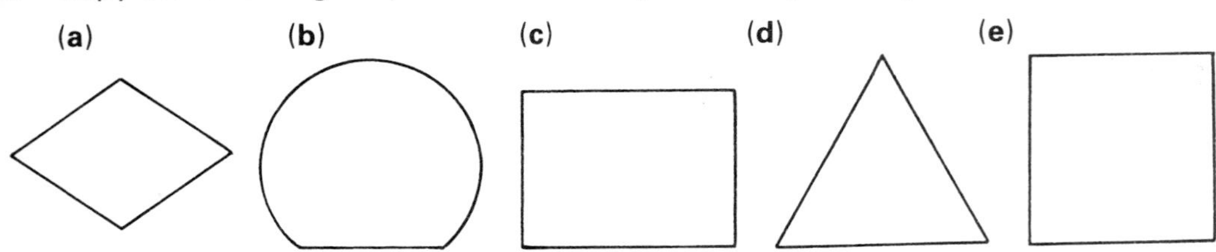

4. Copy the following figures and complete the symmetry:

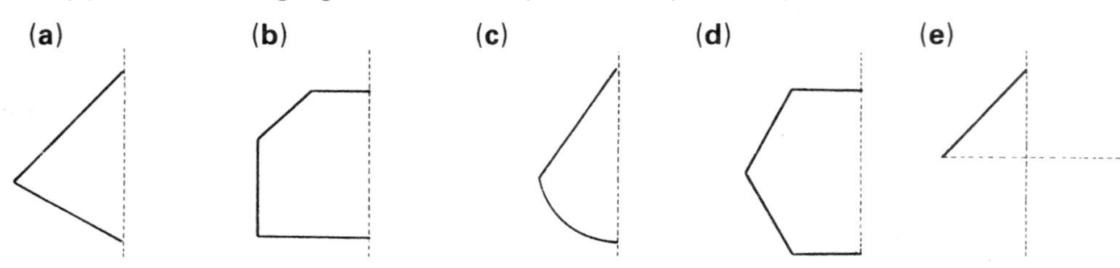

The dotted lines are lines of symmetry in each case.

5. In Question 2 you were able to fold the rectangle along two lines so that one half mapped onto the other half exactly.

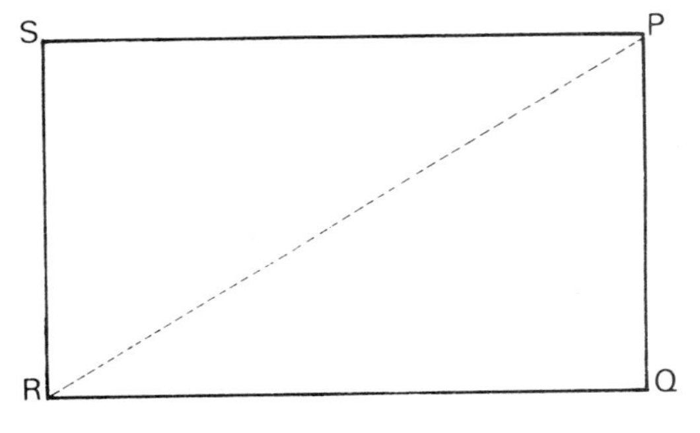

(**a**) If the original corners of the rectangle were lettered PQRS as in the figure on the right, would it be possible to fold the rectangle along the line PR so that Q mapped onto S? Try this and draw a sketch of the result.

(**b**) What does this suggest about the line PR, and about the line QS?

6. Take a rectangular sheet of paper. Measure a length of 3 cm from one corner along one of the longer sides (PB in the diagram). Do the same from the opposite corner (DQ in the diagram) so that PB = DQ and is parallel to it.
Join AB and CD so that you have the shape shaded in the diagram. Now cut along AB and CD to give the shape ABCD.

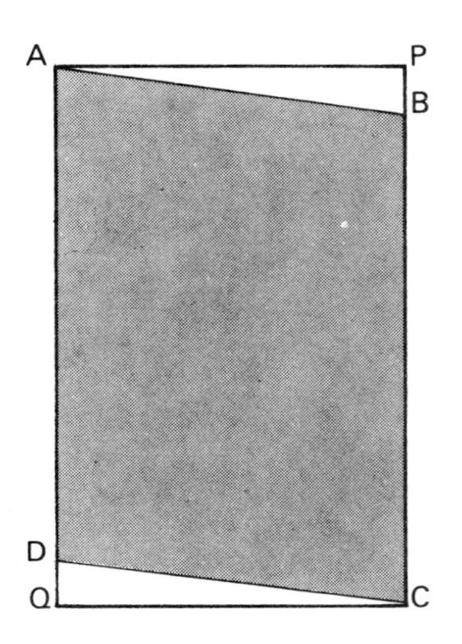

(**a**) What name is given to the shape ABCD?

(**b**) Can you find any lines about which you can fold the shape so that one half maps exactly onto the other; i.e. are there any lines of symmetry?

7. Take another rectangular sheet of paper, this time 16 cm by 8 cm. Repeat what you did in Question 6, this time making PB and DQ 6 cm long. The shape you end up with will be slightly different.

(**a**) What name do we give to the shape ABCD in this case?

(**b**) Are there any lines about which you can fold the figure as in Question 6(b)? Check carefully to see if your results are reasonable.

8. **P Y T H A G O R E A N S**

In the word above, give any letters which have:

(**a**) only a vertical line of symmetry

(**b**) only a horizontal line of symmetry

(**c**) more than one line of symmetry

(**d**) rotational symmetry

(**e**) rotational symmetry only

(**f**) no line or rotational symmetry.

9. On a sheet of plain paper draw a circle of radius 8 cm. Cut this out and fold it so that one half maps exactly onto the other. Open it out and draw in the fold line. (See Fig. 1.) Repeat this to give another line roughly at right angles to the first. (See Fig. 2.) Letter these two lines AB and CD.

Fig. 1

Fig. 2

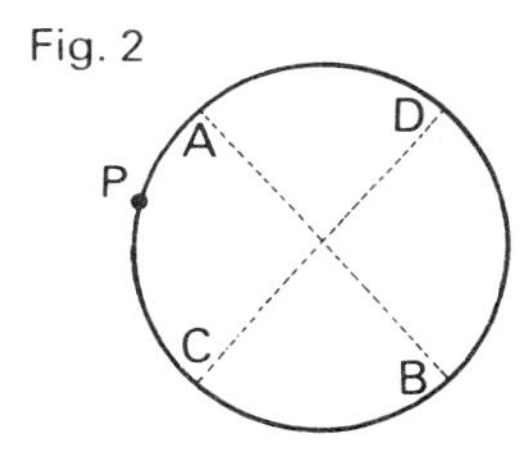

(**a**) What point lies where the two lines of symmetry intersect?

(**b**) How many axes of symmetry does the circle have altogether?

(**c**) Mark a point P on the edge of the circle and note where the image of P is when the paper is folded. Letter this image point P′. Join PP′. What do you notice about PP′ and the line of symmetry?

(**d**) Mark any two points P and P′ on a sheet of plain paper. Use (c) to find quickly the perpendicular bisector of PP′.

10. In the polygon on the right all sides and angles are equal. It is a regular polygon – what is its name? Make a copy of this shape on a sheet of paper by drawing a circle of 5 cm radius. Mark the centre O. Then measure off angles of 72° at O until you have completed one revolution (see the diagram). If necessary extend these lines to cut the circumference of your circle. Join these points around the circumference, i.e. A to B, B to C, . . . , E to A. In this way a regular shape should result.

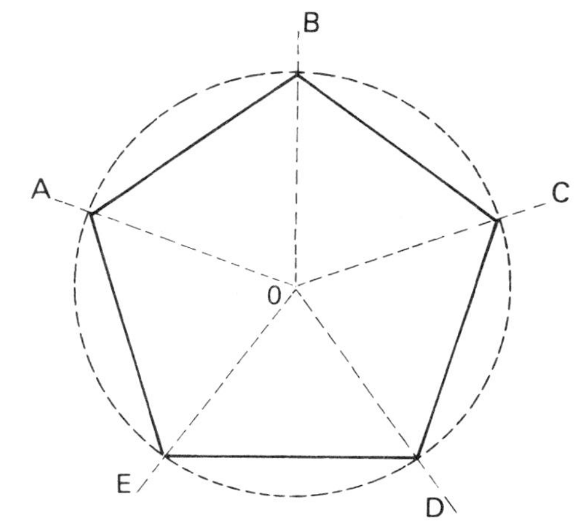

(a) How many lines of symmetry do you find in this case?

(b) What does this suggest about the line symmetry of all regular polygons? (Think of the square and equilateral triangle as well).

(c) What is the rotational symmetry of the figure?

Exercise 17.2 *Reflection*

1. Draw in the x and y axes and number them from -5 to $+5$. Imagine that a double-sided mirror is standing on the line $x = 0$ (make sure that you have the correct axis).
   (a) What would be the coordinates of the image points of A(5,0), B(3,2), C(2,−4), D(0,3), E(4,4) when reflected in this mirror line?
   (b) Draw in the line joining each point to its respective image (A to A′, etc.).
   (c) What do you notice about these lines with respect to the mirror line? (If you have thought carefully you should have noticed two important facts).
   (d) Compare the coordinates of each image with those of the original point (A with A′, etc.). Do you notice anything that is true in every case?
   (e) Remembering that you have been reflecting in the line $x = 0$, does this provide you with a quick way of finding the image coordinates when this particular mirror line is used?
   (f) Are there any points unchanged by this reflection? Why do you think this is?
2. Draw another set of x and y axes numbered from -5 to $+5$. This time use the line $y = 0$ as the mirror line.
   (a) Find the coordinates of the images of points A–E from Question 1(a) in the new mirror line.
   (b) Draw in the line joining each point to its respective image (A to A′, etc.).
   (c) Do you notice that the same facts are true as in Question 1(b)?
   (d) Once again compare the coordinates of each image with those of the original point (A with A′, etc.). Is anything true in every case?
   (e) Does this provide you with a way of remembering what happens to the coordinates when this particular mirror line is used? (See Question 1(e).)
   (f) Are there any points unchanged by this reflection? Why do you think this is?
3. Draw another set of x and y axes numbered from -5 to $+5$. This time use the line $x = y$ as the mirror line.
   (a) Find the coordinates of the image points of A–E from Question 1(a) in the new mirror line.
   (b) Draw in the lines joining each point to its image. Do you notice anything that is true in every case? Is it the same as in the previous questions?
   (c) Once again compare the coordinates of each image point with those of the original point. Is anything true in every case? If so, would it help you to find the image coordinates quickly when this particular mirror line is being used?
   (d) Are there any points unchanged by reflection in this case? Do they fit in with your ideas for Questions 1(f) and 2(f)?
   (e) Explain how you could use the results of part (b) of Questions 1, 2 and 3 to find the image in a mirror line when no coordinates are available to help.

4. Draw in the x and y axes and number them from -6 to $+6$. Take the line $x = 0$ as the mirror line and reflect the points P(−1,1), Q(2,4), R(5,−2). Join P, Q, R to form the triangle PQR and also P′, Q′, R′ to form the image triangle P′Q′R′.
   (a) Write out the coordinates of P′, Q′ and R′. (Question 1(e) might help here).
   (b) Mark points A and B that lie on the perimeters of the triangles but are unchanged by this reflection. Give the coordinates of A and B and explain why they have remained unchanged in position.
   (c) Repeat the question, reflecting the same points P, Q and R in the line $y = 0$. Mark points W and Z which lie on the perimeters of the triangle but are unchanged by this reflection. Give the coordinates of these points as before.
5. Repeat the last question for the triangle PQR but this time take the line $x = 1$ as the mirror line or line of reflection.
   (a) Find the image points P′, Q′ and R′ and join them to form the image triangle P′Q′R′.
   (b) Are there any points unchanged by reflection in this case? Will there generally be two points unchanged by reflection in an x line? Make separate drawings to show any cases where the answer is different.
6. Draw a line AB 10 centimetres long. Mark any point P not on AB. Taking a pair of compasses, draw several circles through P with centres lying on AB. You should find that these circles all intersect at P and at one other point. What connection has this other point with P? Show how this method could be used to construct the image point of any point not on the mirror line. How many circles would you have to draw to find this other point accurately?
7. Write the word 'EXAMS' in block capitals as shown. Draw in the mirror line M_1 (Fig. 1) and reflect the letters in this mirror line. Repeat this using the mirror line M_2 (Fig. 2).
   (a) Is there anything special about the letters that appear unchanged by reflection in each case?
   (b) What can be said about the symmetry of the letters which are unchanged by either reflection?

Fig. 1

EXAMS

M_1

M_2

Fig. 2

EXAMS

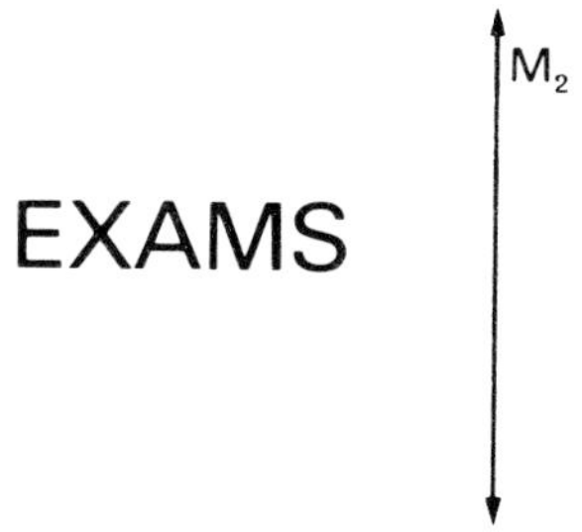

8. A triangle PRQ has vertices P(2,3), Q(5,3), R(4,6).
   (a) Give the coordinates of the image points P′, Q′ and R′ after P, Q, R are reflected in the line $x = 1$.
   (b) Join P′, Q′ and R′ to form the image triangle P′Q′R′. Can PQR be mapped directly onto P′Q′R′ by any other single transformation?
9. The coordinates of the vertices of a triangle T_1 are (2,0), (5,0) and (5,3).
   (a) Reflect T_1 in the line $y = 0$ and draw in its image T_2.
   (b) Reflect T_2 in the line $x = 0$ and draw in its image T_3.
   (c) Reflect T_3 in the line $x = y$ and draw in its image T_4.
   (d) What is the relationship between T_1 and T_4?

10. Reflection in a certain line maps the point (2,2) onto (2,2) and (−5,−2) onto (−2,−5).
   (a) What is the equation of the mirror line? (A drawing might help here.)
   (b) What would be the reflection of the following points in this mirror line?
      (i) A(3,5) (ii) B(0,0) (iii) C(1,−4) (iv) D(−3,4)
11. A certain reflection maps (3,1) onto (−1,1) and (2,5) onto (0,5).
   (a) What is the equation of the mirror line?
   (b) Give the coordinates of the image points of (1,7) and (5,0) under this reflection.
12. P and Q are two houses which have to be connected to the main power cable. In order to save cost they have to be connected to the main cable at the same point and the shortest possible length of cable must be used to do this.

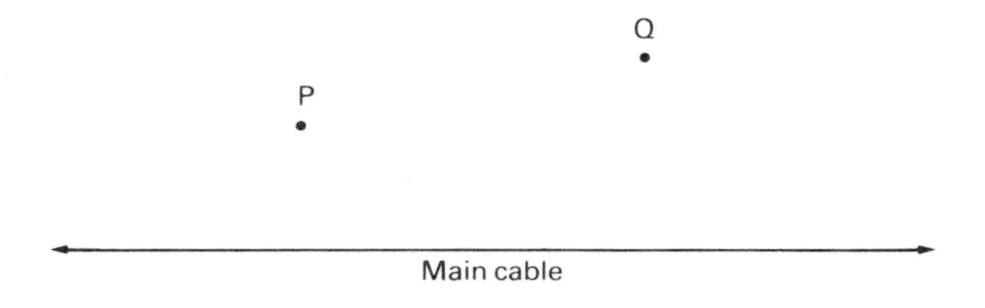

   (a) Try several different positions for the join and the cable. See if reflection could help to show the point at which the cable should be joined.
   (b) Check by measurement.

Exercise 17.3 *Rotation*

1. Draw the x and y axes from −5 to +5 and plot the points A(3,1), B(2,4), C(−2,3), D(−4,−3) and E(2,−2). Using tracing paper trace the axes and these points.
   (a) With your pencil point keep the origin fixed and rotate the tracing paper through an angle of 180° clockwise. This rotation is described as 'a rotation of 180° about the point (0,0)'.
   (b) Mark in the image points A′, B′, C′, D′ and E′ of A, B, C, D, E respectively under this rotation.
   (c) Compare the coordinates of A and A′, B and B′, etc. What do you notice about the change in the coordinates? Would this help you to find the image point of (p,q) under this rotation?
   (d) Join A to O and A′ to O. What do you notice about these two lines? What is the size of angle AOA′? Is this also true for BOB′, etc.?
   (e) Join A to C and A′ to C′. What do you notice about these two lines? Through what angle do you think that AC must be rotated about the origin to map onto A′C′? Is this also true for BD and B′D′, etc.?
   (f) What is the path traced out by A under this rotation? Is this also true for the paths traced out by B, C, D and E?
2. Draw axes as in Question 1 and mark the same points A, B, C, D and E. Use tracing paper as before to trace the axes and points given.
   (a) Rotate the paper through an angle of 90° clockwise about the origin.
   (b) Mark in the image points A″, B″, C″, D″ and E″, the image points of A, B, C, D and E respectively.

(c) Compare the coordinates of A and A″. What do you notice about the change in the coordinates? Would this help you to find the image of the point (a,b) under this rotation?

(d) Describe the rotation as in Question 1(a). Give your answer in the form: 'This rotation is described as "..........".'

(e) Join A to O and A″ to O. What do you notice about the angle AOA″? Is this also true for angle BOB″, etc.?

(f) Join A to C and A″ to C″. What do you notice about the angle between these lines? Through what angle would you rotate AC about the origin to make it map onto A″C″?

(g) What is the path traced out by C under this rotation? What about the paths traced out by A, B, C, D and E?

(h) If they travel on the arcs of circles where do you think the centres will lie?

(i) Can you find one point unchanged in its position by the rotation given? Are there any more?

3. Draw in the axes and points of Question 1. Trace them on tracing paper.

(a) Rotate the tracing paper through an angle of 180° about the point (1,1).

(b) Compare the coordinates of A and A*, its image under this rotation. Are they changed in any special way by this rotation? Is this also true for the coordinates of B, etc.?

(c) Would it have made any difference in this case if the rotation had been clockwise or anticlockwise?

(d) Give one point unchanged in its position by the rotation. Are there any others?

(e) If A, B, C, D and E trace out the arcs of circles under this rotation, where do you think the centres of those circles will lie?

4. Plot the points X(4,0), Y(0,0), Z(0,4) and then join them to form the triangle XYZ.

(a) Rotate the triangle 180° about the origin and mark in the image points X′, Y′, Z′, joining them to form the triangle X′Y′Z′.

(b) What do you notice about XY and X′Y′? Is this the same for XZ and X′Z′?

(c) Dot in a line of reflection that would map triangle XYZ onto triangle Z′Y′X′.

(d) Give the coordinates of one point unchanged in position by this rotation.

(e) If this rotation is called **R**, what would be the effect of doing **R** twice?

(f) Give the mirror lines that would map triangle XYZ onto X′Y′Z′ by two reflections. Are these the only two mirror lines that would map XYZ onto X′Y′Z′?

5. Plot the points A(1,1), B(1,4), C(4,1) to form the triangle ABC.

(a) Rotate the triangle ABC 90° anticlockwise about the point O(0,0). Mark in the image points A″, B″, C″ under this rotation.

(b) What is the angle between the lines AO and A″O, where O is the origin?

(c) Through what angle anticlockwise must BC be rotated to map onto B″C″?

(d) If this rotation is called **Q**, what would be the effect of doing **Q** twice? (This would be written as $\mathbf{Q}^2$.)

(e) What would be the effect of doing $\mathbf{Q}^4$; that is, repeating **Q** four times?

6. Plot the points P(2,2), Q(2,5) and R(5,3) and join them to form the triangle PQR.

(a) Rotate the triangle PQR 180° about the point (2,2) and mark in the image points P*, Q* and R*.

(b) What do you notice about the lines PQ and P*Q* under this rotation?

(c) Does it matter that the rotation in part (a) was not described as clockwise or anticlockwise?

(d) If **H** is the original rotation, what would be the effect of $\mathbf{H}^2$?

7. Plot the points A(0,0), B(2,2), C(4,0), D(2,−2) to form the quadrilateral ABCD.

(a) Rotate ABCD 90° clockwise about the point (0,0) and mark the image points A′, B′, C′, D′ under this rotation.

(b) Which point is unchanged in its position by this rotation?

(c) Letting the original rotation be **T**, mark in A″, B″, C″, D″, the image points of A, B, C, D under the rotation $\mathbf{T}^2$.

(d) Mark in the image points A*, B*, C*, D*, the images after applying $\mathbf{T}^3$ to the square ABCD.

(e) What would you expect to happen after doing $\mathbf{T}^4$ to ABCD?

8. Repeat Question 7, this time taking the centre of rotation as (1,1).

9. In the figure opposite ABC and BCD are equilateral triangles.

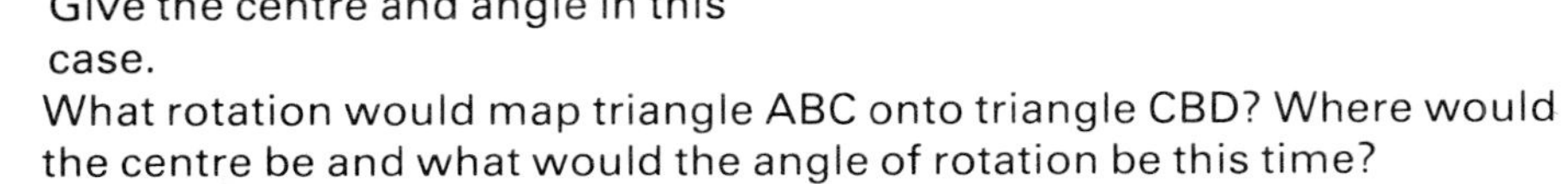

(a) What rotation would map triangle ABC onto triangle BDC? (This implies A → B, B → D, C → C.) Where would the centre be? What would the angle of rotation be?

(b) What rotation would map triangle ABC onto triangle DCB? Give the centre and angle in this case.

(c) What rotation would map triangle ABC onto triangle CBD? Where would the centre be and what would the angle of rotation be this time?

10. How many different centres of rotation could be found to map the square ABCD, with coordinates A(−5,5), B(1,5), C(1,−1), D(−5,−1), onto the square PQRS with coordinates P(−1,1), Q(5,1), R(5,−5), S(−1,−5)? (Do not take the lettering in this case to imply a mapping. Any point of ABCD can be mapped onto any point of PQRS.) Give the centre and the angle in each case.

18

Further arithmetic

Exercise 18.1 *Order of operations with whole numbers*

Work out the following:

1. $(9-4)\times 2$
2. $9-(4\times 2)$
3. $6+(3\times 2)$
4. $(6+3)\times 2$
5. $(14\div 7)+5$
6. $14\div(7-5)$
7. $14+(3\times 2)-8$
8. $(14+3)\times(8-2)$
9. $(27\div 3)+(6\times 4)$
10. $(9+7-3)\times 5$
11. $9+7-(3\times 5)$
12. $(5\times 4\times 3)-2$
13. $(5\times 4)\times(3-2)$
14. $(24\div 6)\times(7-3)$
15. $27\div(3+6)\times 4$
16. $16+2\times 4$
17. $16\div 2-4$
18. $3\times 7-6\div 3$
19. $3+18\div 6-2$
20. $3+18-6\div 2$
21. $18\div 3-6+2$
22. $10\times 14\div 7+2$
23. $10+14\div 7-2$
24. $4\times 6+10\div 2$
25. $4+6\div 2+10$

Exercise 18.2 *Brackets and fractions*

Simplify the following:

1. $(\frac{1}{2}+\frac{1}{3})\times\frac{9}{10}$
2. $(\frac{3}{5}\div\frac{3}{4})-\frac{1}{2}$
3. $\frac{5}{6}+(\frac{5}{8}\times\frac{14}{15})$
4. $\frac{2}{3}\div(\frac{3}{4}-\frac{1}{6})$
5. $(3\frac{1}{3}+2\frac{1}{2})\times 1\frac{2}{7}$
6. $(3\frac{1}{3}-2\frac{1}{2})\div 1\frac{1}{4}$
7. $4\frac{3}{4}+(2\frac{1}{2}\times 1\frac{1}{5})$
8. $(3\frac{1}{3}\times 2\frac{1}{2})+2\frac{1}{2}$
9. $(3\frac{1}{3}\div 2\frac{1}{2})-\frac{3}{4}$
10. $3-(2\frac{1}{4}\div 1\frac{1}{5})$
11. $3\frac{3}{7}\times(2\frac{1}{3}+1\frac{3}{4})$
12. $\frac{2}{3}$ of $(2\frac{1}{3}-1\frac{3}{4})$
13. $2\frac{4}{5}\div(4\frac{2}{3}-1\frac{2}{5})$
14. $4\frac{1}{3}-(2\frac{2}{5}\times 1\frac{3}{7})$
15. $(\frac{1}{2}+\frac{1}{3})\div(\frac{1}{2}-\frac{1}{3})$
16. $(\frac{3}{4}+\frac{1}{2})\times(\frac{3}{4}-\frac{1}{2})$
17. $(\frac{3}{4}\times\frac{1}{2})+(\frac{3}{4}\div\frac{1}{2})$
18. $(\frac{7}{10}\div\frac{3}{5})-(\frac{3}{16}\times\frac{8}{9})$
19. $(3\frac{1}{7}\times 2\frac{4}{5})-(4\frac{2}{3}\div 1\frac{1}{9})$
20. $(2\frac{1}{4}+\frac{19}{20})\times(1\frac{1}{2}-\frac{5}{8})$
21. $(\frac{3}{5}$ of $3\frac{3}{4})+(2\frac{4}{9}\div 1\frac{5}{6})$
22. $(1\frac{2}{3}+1\frac{5}{6})\div(5-2\frac{3}{8})$
23. $(1\frac{2}{3}-\frac{11}{12})\times(\frac{11}{14}+1\frac{1}{2})$
24. $(2\frac{4}{5}\times 3\frac{3}{4})-(2\frac{4}{9}\div 3\frac{2}{3})$
25. $(1\frac{1}{2})^3+(\frac{15}{16}$ of $2\frac{2}{3})$

Exercise 18.3 *Combined operations*

Simplify the following:

1. $3\frac{1}{2}\div 2\frac{5}{8}+1\frac{1}{2}$
2. $3\frac{1}{2}+2\frac{5}{8}\div 1\frac{1}{2}$
3. $6\frac{1}{4}\times 1\frac{3}{5}-2\frac{1}{3}$
4. $\frac{3}{4}$ of $1\frac{7}{9}+1\frac{11}{12}$
5. $6\frac{9}{10}+1\frac{3}{7}\times 1\frac{3}{4}$
6. $2\frac{2}{3}+2\frac{1}{3}\div 1\frac{5}{9}$
7. $6-\frac{2}{3}$ of $1\frac{2}{7}$
8. $\frac{3}{4}$ of $3\frac{1}{6}-\frac{5}{6}$
9. $5\frac{1}{4}\times 1\frac{2}{3}-5\frac{5}{6}$
10. $4\frac{1}{3}-2\frac{5}{6}\div 1\frac{3}{8}$
11. $\frac{3}{4}\times\frac{2}{3}+\frac{1}{2}\div\frac{4}{5}$
12. $\frac{3}{4}\div\frac{2}{3}-\frac{1}{2}\times\frac{4}{5}$
13. $\frac{3}{4}+\frac{2}{3}\div\frac{1}{2}-\frac{4}{5}$
14. $\frac{3}{4}$ of $\frac{2}{3}+\frac{1}{2}-\frac{4}{5}$
15. $\frac{3}{4}+\frac{2}{3}-\frac{1}{2}$ of $\frac{4}{5}$
16. $1\frac{3}{4}\div 1\frac{1}{2}+2\frac{1}{3}\times 2\frac{1}{2}$
17. $1\frac{1}{2}\times 1\frac{1}{6}-1\frac{1}{2}\div 1\frac{1}{6}$
18. $1\frac{1}{9}\times 1\frac{1}{2}+\frac{4}{5}-1\frac{1}{2}$

19. $2\frac{1}{6} - \frac{2}{3} - \frac{2}{3} \div 1\frac{1}{2}$

20. $2\frac{1}{4} + 3\frac{1}{3} \times 1\frac{1}{5} - 3\frac{2}{3}$

Exercise 18.4 *Long division – whole numbers*

Do the following divisions:

1. $851 \div 23$
2. $1426 \div 31$
3. $1073 \div 37$
4. $1927 \div 41$
5. $3901 \div 47$
6. $3363 \div 57$
7. $2135 \div 61$
8. $2184 \div 39$
9. $4732 \div 52$
10. $1292 \div 19$
11. $4238 \div 13$
12. $2852 \div 23$
13. $2193 \div 17$
14. $7223 \div 31$
15. $6956 \div 47$
16. $10\,101 \div 37$
17. $23\,664 \div 29$
18. $51\,663 \div 51$
19. $11\,739 \div 43$
20. $17\,082 \div 73$

Exercise 18.5 *Long division – decimals*

Divide:

1. 20.88 by 3.6
2. 22.05 by 4.5
3. 19.71 by 0.27
4. 1.512 by 1.8
5. 4.655 by 0.49
6. 0.1064 by 0.028
7. 1.072 by 1.6
8. 24.192 by 6.3
9. 16.992 by 0.72
10. 7.854 by 0.014
11. 15.87 by 2.3
12. 0.2805 by 0.17
13. 251.74 by 4.1
14. 0.4161 by 0.73
15. 0.020 02 by 0.013
16. 1458.6 by 3.9
17. 24.589 by 0.67
18. 0.2015 by 0.031
19. 27.058 by 0.83
20. 17.907 by 4.7

In each of the following calculations, use the answer to (a) to find the value of parts (b) and (c):

21. (a) $8.688 \div 2.4$ (b) $86.88 \div 2.4$ (c) $0.8688 \div 0.24$
22. (a) $137.92 \div 3.2$ (b) $1.3792 \div 0.32$ (c) $1.3792 \div 32$
23. (a) $0.1533 \div 0.21$ (b) $1.533 \div 2.1$ (c) $153.3 \div 0.21$
24. (a) $122.12 \div 4.3$ (b) $122.12 \div 0.43$ (c) $1.2212 \div 0.043$
25. (a) $0.9996 \div 0.51$ (b) $9.996 \div 51$ (c) $0.099\,96 \div 5.1$

Exercise 18.6 *Combined operations with decimals*

Work out the following:

1. $\dfrac{4.3 + 1.7}{1.2}$
2. $\dfrac{6 - 3.3}{3}$
3. $\dfrac{(0.2)^2 + 1.6}{4}$
4. $\dfrac{11.95 - 3.2}{0.5}$
5. $\dfrac{4.2 \times 1.6}{0.7}$
6. $\dfrac{1.44 \times 0.4}{0.08}$
7. $\dfrac{30}{1.2 \times 0.5}$
8. $\dfrac{17.6 \times 4.8}{80}$
9. $(14.3 + 2.9) \times 0.6$
10. $14.3 + 2.9 \times 0.6$
11. $17.2 \div 0.4 - 11.7$
12. $(22.5 \div 0.3) - (22.5 \times 0.3)$
13. $0.8 + 0.7 \times 0.5$
14. $0.8 \times 0.7 + 0.5$
15. $3.8 - (1.7 \times 0.3) + 2.75$
16. $(0.6)^2 - 0.018 \div 0.3$
17. $(80 - 2.4) \div 0.02$
18. $2.4 \times 0.6 + 2.4 \div 0.6$
19. $\dfrac{9.6 \times 1.5}{3 \times 4.8}$
20. $\dfrac{10.8 \times 1.25}{2.7 \times 0.5}$

19

Percentage

Exercise 19.1 *Relationship to fractions and decimals*

Write as fractions in lowest terms:

1. 17%	2. 10%	3. 25%	4. 40%	5. 35%
6. 50%	7. 6%	8. 75%	9. 8%	10. 85%
11. 24%	12. 60%	13. 84%	14. 5%	15. $33\frac{1}{3}\%$
16. $66\frac{2}{3}\%$	17. $16\frac{2}{3}\%$	18. $2\frac{1}{2}\%$	19. $8\frac{1}{3}\%$	20. $17\frac{1}{2}\%$

Write as percentages:

21. $\frac{1}{100}$	22. $\frac{1}{50}$	23. $\frac{1}{25}$	24. $\frac{1}{20}$	25. $\frac{1}{10}$
26. $\frac{1}{5}$	27. $\frac{1}{4}$	28. $\frac{1}{2}$	29. $\frac{6}{25}$	30. $\frac{3}{5}$
31. $\frac{61}{100}$	32. $\frac{3}{4}$	33. $\frac{9}{10}$	34. $\frac{17}{50}$	35. $\frac{13}{20}$
36. $\frac{1}{8}$	37. $\frac{1}{6}$	38. $\frac{1}{3}$	39. $\frac{1}{40}$	40. $\frac{2}{3}$

Write as decimals:

41. 23%	42. 40%	43. 38%	44. 25%	45. 16%
46. 50%	47. 87%	48. 75%	49. 5%	50. $37\frac{1}{2}\%$
51. $14\frac{1}{4}\%$	52. $6\frac{1}{2}\%$	53. $32\frac{3}{4}\%$	54. $12\frac{1}{2}\%$	55. $2\frac{1}{2}\%$

Write as percentages:

56. 0.47	57. 0.86	58. 0.07	59. 0.6	60. 0.01
61. 0.5	62. 0.025	63. 0.375	64. 0.3	65. $0.\dot{3}$

Arrange in ascending order (start with the smallest):

66. $\frac{7}{20}$, 33%, 0.34	67. $\frac{4}{25}$, 0.15, 17%	68. $\frac{1}{3}$, 31%, 0.3
69. 64%, $\frac{13}{20}$, 0.6	70. $\frac{2}{3}$, 0.67, $\frac{33}{50}$, 65%	71. $\frac{4}{9}$, 0.45, 43%, $\frac{11}{25}$

Exercise 19.2 *Percentage calculations*

Work out each of these:

1. 23% of 100	2. 19% of 200	3. 75% of 40	4. 50% of 11
5. 6% of 25	6. $12\frac{1}{2}\%$ of 80	7. $6\frac{1}{4}\%$ of 48	8. $33\frac{1}{3}\%$ of 120
9. 25% of 46	10. $2\frac{1}{2}\%$ of 160	11. 20% of £1	12. 25% of £2
13. 18% of £4	14. 44% of £5	15. 9% of £20	16. $66\frac{2}{3}\%$ of £3
17. $2\frac{1}{2}\%$ of £2	18. $33\frac{1}{3}\%$ of £45	19. $6\frac{1}{4}\%$ of £80	20. $\frac{1}{2}\%$ of £8

21. 80% of £2.50
22. 35% of £1.60
23. 5% of £3.40
24. 8% of £1.25
25. 18% of £9.50
26. $7\frac{1}{2}$% of £26
27. 7% of 1 m
28. $12\frac{1}{2}$% of 4 m
29. 60% of 1 cm
30. 19% of 5 cm
31. $2\frac{1}{2}$% of 1 km
32. 36% of 10 km
33. 45% of 1 g
34. 75% of 20 g
35. 10% of 1 kg
36. 18% of 6 kg
37. 80% of 1 tonne
38. $37\frac{1}{2}$% of 2 tonnes
39. 84% of 3 litres
40. $7\frac{1}{2}$% of 2 litres

Exercise 19.3 *One quantity as a percentage of another*

Express the first quantity as a percentage of the second:

1. 6, 24
2. 17, 50
3. 23, 25
4. 11, 20
5. 9, 10
6. 45, 60
7. 16, 128
8. 60, 180
9. 12, 480
10. 70, 105
11. $7\frac{1}{2}$, 45
12. $8\frac{1}{4}$, 33
13. $3\frac{1}{2}$, $17\frac{1}{2}$
14. $2\frac{1}{4}$, 18
15. 16p, £1
16. 48p, £4
17. 80p, £5
18. 15p, £2
19. 25p, £1.25
20. £2.70, £3.60
21. £1.50, £2.50
22. £3.15, £7.50
23. 30p, £2.40
24. 42p, £1.26
25. 8p, £1.50
26. 12p, 75p
27. 6p, 20p
28. 4p, 12p
29. 36 cm, 1 m
30. 8 mm, 5 cm
31. 160 m, 1 km
32. 400 g, 1 kg
33. 250 ml, 2 litres
34. 400 kg, 1 tonne
35. 500 m, 4 km
36. 25 kg, 1 tonne
37. 15 mm, 6 cm
38. 700 g, 4 kg
39. 80 cm, 2 m
40. 8 cm, 4 m

Exercise 19.4 *Problems*

1. Hilary got 36 out of 80 for a test. What percentage is this?
2. An examination is marked out of 70. Pupils who gain 60% will pass. What is the pass mark?
3. A family spends £600 on holiday. 45% is spent on accommodation, 20% on travel and the rest on entertainment. How much is spent on each item?
4. 80% of a class of 30 like pop music. How many do not like it?
5. A boy spends only $33\frac{1}{3}$% of the lesson working. If the lesson is three quarters of an hour long, for how many minutes does he work?
6. 40% of the children in a school of 600 live within 2 km of their school. How many live more than 2 km away?
7. Income tax is collected at a rate of 30%. What does a man pay if he is to be taxed on £1150 at this rate?
8. Add together 15% of 60 and 60% of 15. What percentage of 25 is this total?
9. A shopkeeper buys a shirt for £4. He sells it to make a profit of 30% on the cost price. What is his profit?
10. A car salesman is given 15% of the sale price of a car as his commission. How much does he earn if he sells a car for £3750?
11. 75% of the children in a school of 240 have had measles. How many have not had measles?
12. A man invests £150 at a rate of 7% interest per annum. How much interest does he receive in a year?
13. A woman earns £80 per week. She is given a rise of £6. What is the percentage increase?

14. At the General Election the total electorate in a constituency was 52 000. The winning candidate got 42% of the possible votes. How many votes did he get?

15. Which is larger, (100 + 10% of 100) or (120 − 10% of 120)?

16. A craftsman produces two dozen tables each week. He increases production by $12\frac{1}{2}$%. What is his new weekly production figure?

17. A woman buys a house for £18 000 and then spends 35% of this total on repairs and renewals. What does the house cost her altogether?

18. A gardener knows that 6% of the bulbs he plants will not bloom. If he plants 750 bulbs how many will bloom?

19. A man is awarded an 8% salary increase. What is his new salary if he used to earn £6500 per annum?

20. After a 15% pay rise a woman earns £92 per week. What was her weekly wage before the increase?

21. In a sale goods are reduced by 20%.

(**a**) A table used to cost £90. What is the sale price?

(**b**) A chair costs £40 in the sale. What is its normal price?

22. A car depreciates in value by 20% of its value at the beginning of the year. If it cost £3200 new, what is its value after (**a**) 1 year (**b**) 2 years?

23. A shopkeeper allows a customer 5% discount for cash settlement. If the customer pays £76 cash, what was the original bill?

24. A housewife pays £180 for an electric stove originally marked at £200. What percentage discount is she given?

25. (**a**) What is 120 increased by 50%?

(**b**) What number when increased by 50% is 120?

26. A joint of meat weighs 2.5 kg. During cooking its real weight is reduced by 15%. What is its weight after cooking?

27. During 1980 the population of a town rose by $2\frac{1}{2}$%. What was the population in January 1981 if it was 54 000 in January 1980?

28. A local councillor is elected with a total of 1544 votes. At the next election there is a swing of $12\frac{1}{2}$% against her. How many people voted for her on this occasion?

20

Ratio

Exercise 20.1 *Simple ratio*

Express each of the following pairs in the form of a ratio $a:b$. Then cancel the ratio down if it is possible. (Remember that in a ratio both parts must be in the same unit and should be whole numbers.)

1. 2 metres, 3 metres
2. 4 cm, 6 cm
3. 6, 36
4. 100, 88
5. 6 pence, 13 pence
6. 32 pence, 96 pence
7. 32 pence, 1 pound
8. 100 metres, 1 km
9. 7 km/h, 17 km/h
10. 63 mg, 1.2 gram
11. 1 cm, 1 metre
12. 1 square centimetre, 1 square metre
13. 25 square centimetres, 0.1 square metre
14. 1 cubic centimetre, 1 litre
15. 3 years 6 months, 4 years 1 month
16. 35 minutes, 65 minutes
17. 3.5, 6.5
18. $\frac{1}{2}$, $\frac{1}{3}$
19. 6 pigs, 3 sheep (careful here!)
20. The number of pigs, the number of sheep (use the numbers in Question 19)

Exercise 20.2 *Problems*

1. A school contains 60 girls and 90 boys.
   (**a**) What is the ratio of the number of girls to the number of boys?
   (**b**) What is the ratio of the number of boys to the total number of pupils?
   (**c**) What fraction of the whole school are boys?
2. The sides of two equilateral triangles are 4 cm and 6 cm respectively. Find the ratio of (**a**) their sides (**b**) their perimeters.
3. The sides of two squares are respectively 6 cm and 8 cm. Find the ratio of (**a**) their perimeters (**b**) their areas.
   Do you notice anything special about your answer to (b) compared to that for (a)?
4. A train travels 150 km in two hours, and a car travels 90 km in an hour. Write down the ratio of the speed of the train to that of the car.
5. A straight line is divided into two equal parts. What is the ratio of (**a**) one part to the other (**b**) one part to the whole line?
6. The interior angles of a regular pentagon are each 108° and those of a regular hexagon are each 120°.
   (**a**) Find the ratio of these interior angles.
   (**b**) Calculate the exterior angle in each case and work out their ratio.
7. You sleep 10 hours a day. What is the ratio of the time you sleep to the time that you are awake in (**a**) a week (**b**) a year?
8. A car travels at 80 km/h and a cyclist travels at 24 km/h. What is the ratio of
   (**a**) their speeds in km/h

(b) their speeds in metres/second? (Think carefully!)

9. Two bicycles cost £75 and £105 respectively when new. After one year their value had dropped by £27 in both cases. What was the ratio of their values (a) when they were bought (b) after one year?

10. The mass of a certain substance is 3 g for 10 cm^3. Another substance has a mass of 0.03 kg for 0.1 litres.
(a) What is the ratio of their masses?
(b) What is the ratio of their volumes?

Exercise 20.3 *Ratios in the form 1 : x*

Express the following ratios in the form 1 : x (example: 2 : 4 would be 1 : 2.)
Note: x may be a fraction or decimal in some cases.

1. 4 : 8
2. 3 : 12
3. 7 : 21
4. 5 cm : 25 cm
5. 3 cm : 5 cm
6. 10 cm : 13 cm
7. 6 metres : 15 metres
8. 0.6 metres : 1.5 metres
9. 1 kg : 0.6 kg
10. 0.6 kg : 1 kg

Exercise 20.4 *Ratios in the form x : 1*

Now repeat Exercise 20.3, except that the ratios should be expressed in the form x : 1 (example: 2 : 4 would be expressed as $\frac{2}{4}$: 1, simplified to $\frac{1}{2}$: 1 or 0.5 : 1).

Exercise 20.5 *Equivalent ratios*

Find x for the following pairs of equal ratios:

1. $2 : 4 = x : 1$
2. $3 : 5 = x : 1$
3. $10 : 15 = x : 6$
4. 60 pence: 50 pence $= x : 1$
5. $7 : 4 = x : 144$
6. $3 : 5 = x : 105$
7. $2 : 3 = x$: 45 pence
8. $2 : 3 = 40 : x$
9. $5 : 6 = 35$ cm : x cm
10. $36 : 42 = 12$ pence : x pence

Exercise 20.6 *Further problems*

1. (a) If 6 bars of chocolate cost 72 pence, how much will 10 bars cost?
(b) What is the ratio of the numbers of bars?
(c) What is the ratio of the costs?
(d) What do you notice?
2. (a) If 6 model cars cost 72 pence, what would be the cost ratio if I bought 10 cars?
(b) Using this, find how much 10 cars would cost.
3. (a) 6 metres of ribbon cost 72 pence. How much would 2 metres cost?
(b) What is the ratio of the amounts of ribbon?
(c) What is the ratio of the costs?
(d) What do you notice?
(e) Use the ideas of parts (a)–(d) to find the cost of 5 metres using ratios.
4. A car travels for 6 hours at 50 km/h. How long would the same journey take at 75 km/h? (Try to use the idea of ratios.)
5. 3 kg of flour cost 63 pence. What quantity of flour would cost 105 pence?

6. It takes 6 men 12 days to weed a field. How long would it take 9 men to weed the same field? (Think carefully!)
7. The ratio of the circumferences of two wheels is 162 : 126. In a certain journey the larger wheel makes 315 revolutions; how many does the smaller one make?
8. A candle is 16 cm long. After burning for 40 minutes it is 14.5 cm long. After how many more minutes will it be 2.5 cm long?
9. A school has enough food for 24 days if everyone eats the normal amount. It is isolated because of a blizzard. How long will the food last if everyone's ration is reduced in the ratio 4 : 5?
10. On a map with a scale of 2 cm to 1 km, the distance between two towns is 2.4 cm. What would be the distance between them on a map with a scale of 3 cm to 1 km?

Exercise 20.7 *Proportion*

1. Draw a line 8 cm long. Letter it AB. Mark a point P 5 cm from A.
   (a) What is the length of PB?
   (b) What fraction of the whole line is AP?
   (c) What fraction of the whole line is PB?
   (d) What is the ratio AP : PB?
2. Divide 24 sweets in the ratio 3 : 5.
   (a) What fraction of the whole is the first number of the ratio?
   (b) What must your two parts add up to altogether?
3. Divide 360° in the ratio 4 : 5.
   (a) What angles do you get for the two parts of the ratio?
   (b) What is the ratio of the smaller angle to the larger?
4. Three boys of ages 3, 4 and 5 years respectively are to divide 60 pence in the ratio of their ages. How much will each boy get?
5. Divide 20 cm in the ratio 2 : 3 : 5. What fraction of the whole will the first part be?
6. Divide a line 15 cm long so that one part is half as long again as the other.
7. The sides of a triangle are in the ratio 2 : 3 : 4 and the perimeter is 18 cm. Find the length of each side.
8. The sides of a triangle are in the ratio 3 : 4 : 5. Find the length of each side if its perimeter is 36 cm.
9. Two sides of a rectangle are in the ratio 3 : 4. What are the lengths of the two sides if its area is 48 cm^2?
10. The ratio of the areas of two squares is 4 : 9. Does this tell you anything about the lengths of the sides of the two squares? Give reasons for your answer.

Exercise 20.8 *Similarity*

1. Draw a square ABCD with sides 3 cm long. Mark a point O where the two diagonals of the square meet. Measure OA and then produce (extend) OA to A′ where $AA' = OA$. (The figure at the top of the next page gives you some idea of what you should have by the time that you have also produced OB, OC and OD in the same way.)
   When you have produced (extended) OB to B′ so that $BB' = OB$ and then done the same for OC and OD, then join A′B′C′D′. This should be another square.

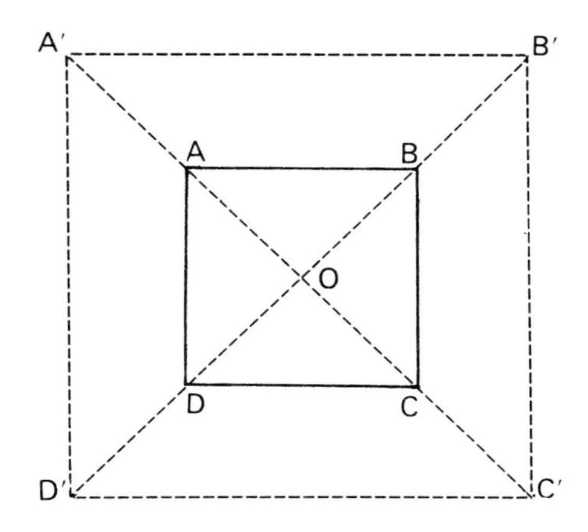

(**a**) What is the ratio OA : OA′?
(**b**) What is the ratio AB : A′B′?
(**c**) Is the same true of all the other sides? Check to see.

2. Repeat Question 1 for the same sized square, but this time produce OA to A′ so that AA′ = 2OA.
(**a**) Answer parts (a) to (c) of Question 1.
(**b**) Supposing that you had produced OA such that AA′ = 3OA, what do you think would have been the ratios for OA : OA′ and AB : A′B′?

3. Make a copy of the triangle ABC, keeping it well to the left-hand side of the page.

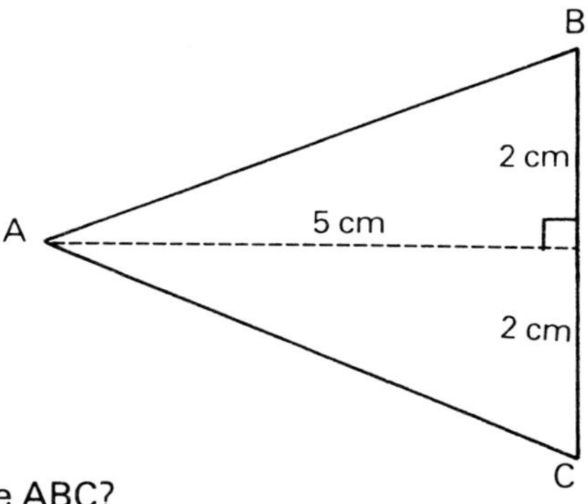

(**a**) Produce AB to B′ so that AB = BB′.
(**b**) What is the ratio AB : AB′?
(**c**) Produce AC to C′ in a similar way such that AC = CC′.
(**d**) What is the ratio AC : AC′?
(**e**) Join B′ to C′. What is the ratio of BC to B′C′?
(**f**) Which is the corresponding angle to angle ABC?
(**g**) What does this suggest about BC and B′C′?
(**h**) Are the triangles ABC and AB′C′ similar?

4. Repeat Question 3, but this time extend AB to B′ so that BB′ = 2AB. Do the same for AC.
(**a**) Answer part (b) and parts (d) to (h) of Question 3.
(**b**) What do you think your results would have been if you had produced AB to B′ such that BB′ = 3AB and also CC′ = 3AC? Give reasons for your answers. If you are not sure, try it out.

5. (**a**) Draw a square with sides 3 cm long. Mark a point O inside the square but not at its centre. Now answer Question 2 for your square and the new O.
(**b**) Draw another square ABCD with sides 3 cm long. Mark a point O outside the square. Now answer Question 2 for this square and your new point O. (You may have to redraw the figure a few times to get everything on the page. Think how best to do this.)

6. Now try the same idea with the triangle ABC. Place the point from which you wish to start all your lines (like O in question 5) in various positions, then try to enlarge the triangle ABC by the same method as before. (Produce OA to A′ so that OA = AA′ or so that AA′ = 2OA etc.) Check that the 'image' triangle A′B′C′ is always similar to the triangle ABC.

7. In the figure on the right AB is parallel to A′B′.

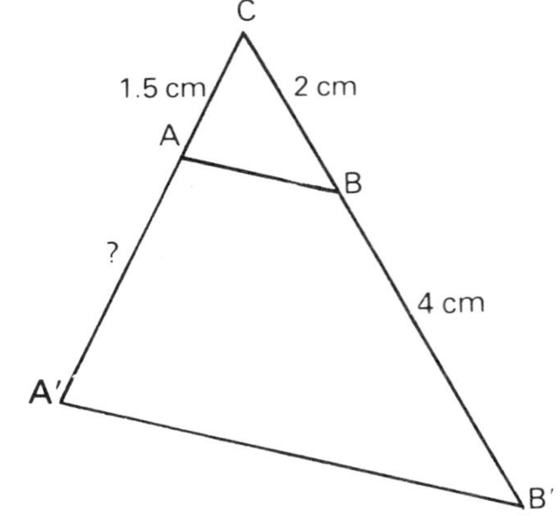

(a) Is triangle ABC similar to triangle A′B′C? Give reasons.
(b) What is the ratio of BC to B′C?
(c) Which point would you call the centre of enlargement?
(d) Do any other points stay in their original position under this enlargement (triangle ABC is enlarged to triangle A′B′C)?
(e) Reduce the ratio B′C to BC to the form $x : 1$ (x is called the scale factor or enlargement factor). In this question x is a whole number.
(f) Multiply AB by the scale factor. What length is this equal to?
(g) Calculate the length of AA′.

8. Give the scale factors of the enlargements in Questions 1, 2, 2(b), 3, 4, 4(b) and 7.
(Remember that the scale factor must be expressed as a number x. However, x may not be a whole number and could even be less than 1.)
In Question 1, suppose that you started with the square A′B′C′D′ and then went halfway toward the point O. What would be the ratio OA : OA′ when expressed in the form $x : 1$?

9.

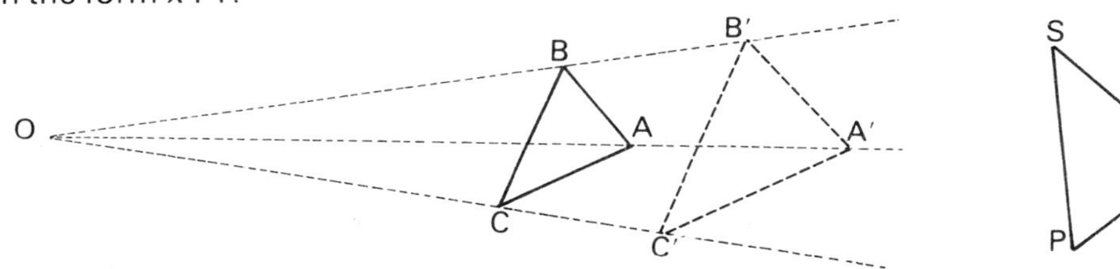

In the drawing above, O is the centre of enlargement to enlarge triangle ABC onto triangle A′B′C′. PRS is a triangle similar to triangle A′B′C′ such that A′B′ : PR = 1 : 1.
(a) Is triangle ABC similar to triangle A′B′C′?
(b) Is triangle A′B′C′ similar to triangle PRS? Do we give this kind of similarity a special name?
(c) O is the centre of enlargement to map ABC onto A′B′C′. Is it also the centre of enlargement to map ABC onto PRS? Give good reasons for your answer.
(d) What is the scale factor if OA = 4 cm and OA′ = 6 cm?
(e) If BC = 2 cm what will be the length of B′C′?
(f) What then would be the ratio RS : B′C′?

10. The pairs of triangles on page 95 are similar. You have to work out the scale factor in each case and find the length of the unknown sides. For convenience we have lettered the vertices ABC and PQR in every case.

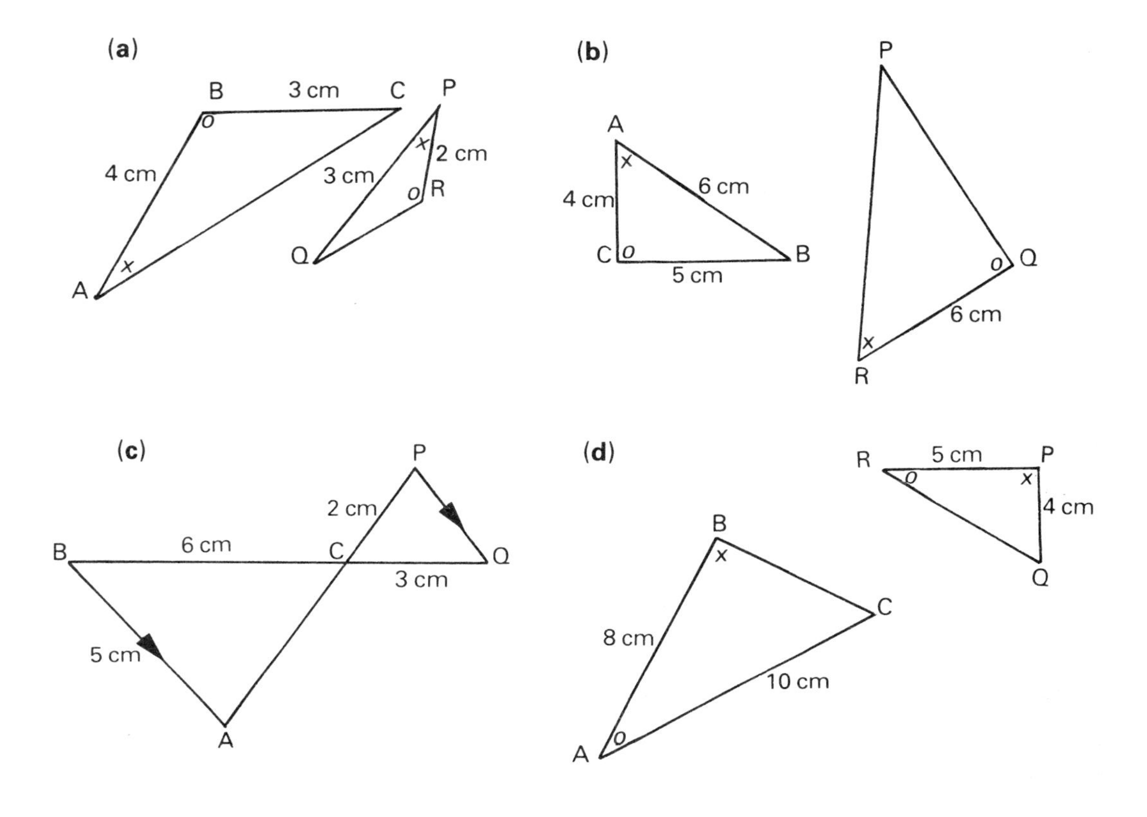

Exercise 20.9 *Increasing and decreasing in ratio*

Increase the following in the ratio given:

1.	6, 3 : 1	**2.**	13, 4 : 1	**3.**	36, 4 : 3	4.	7.5, 5 : 3
5.	25p, 7 : 5	**6.**	120 cm, 11 : 8	**7.**	17, 3 : 2	8.	15 kg, 7 : 4
9.	£1.64, 5 : 4	**10.**	65 cm^2, 13 : 10				

Decrease the following in the ratio given:

11.	9, 2 : 3	**12.**	25, 3 : 5	**13.**	72, 8 : 9	**14.**	20.5, 23 : 41
15.	65p, 7 : 13	**16.**	2.4 kg, 3 : 8	**17.**	12.5, 4 : 50	**18.**	£2.75, 3 : 25

Exercise 20.10 *Finding the ratio of increase or decrease*

Find the value of x in the following ratios:

1. 12 is increased to 36 in the ratio $x : 2$.
2. 25 is decreased to 20 in the ratio $x : 5$.
3. 91 is decreased to 65 in the ratio $5 : x$.
4. 17 is increased to 68 in the ratio $4 : x$.
5. 121 is decreased to 66 in the ratio $6 : x$.
6. 14.4 cm is increased to 20.4 cm in the ratio $x : 12$.
7. 13.2 kg is decreased to 3.3 kg in the ratio $x : 6$.
8. 25 cm is decreased to 6.25 cm in the ratio $2 : x$.
9. 27 is increased to 94.5 in the ratio $x : 2$.

21

Degree of accuracy and standard form

Exercise 21.1 *Nearest whole number*

Write to the nearest whole number:

1. 4.2 **2.** 4.6 **3.** 13.43 **4.** 24.96 **5.** 6.49
6. 6.51 **7.** 6.5 **8.** 9.99 **9.** 0.73 **10.** 0.5

Exercise 21.2 *Decimal places*

Write these correct to 1 decimal place:

1. 3.67 **2.** 4.93 **3.** 12.86 **4.** 24.03 **5.** 1.54
6. 0.56 **7.** 6.55 **8.** 8.98 **9.** 9.96 **10.** 89.95

Write these correct to 2 decimal places:

11. 6.734 **12.** 5.879 **13.** 17.481 **14.** 36.444 **15.** 0.069
16. 0.094 **17.** 0.097 **18.** 0.006 **19.** 0.195 **20.** 8.999

Write these correct to 3 decimal places:

21. 4.3729 **22.** 1.8642 **23.** 7.7487 **24.** 5.1264 **25.** 4.3107
26. 2.4702 **27.** 1.9896 **28.** 2.4365 **29.** 0.0098 **30.** 9.9999

Exercise 21.3 *Significant figures*

Write these correct to 1 significant figure:

1. 37 **2.** 43 **3.** 65 **4.** 1047 **5.** 7501
6. 9876 **7.** 84 998 **8.** 170 000 **9.** 14 970 **10.** 97 429

Write these correct to 2 significant figures:

11. 375 **12.** 208 **13.** 195 **14.** 3046 **15.** 9870
16. 70 500 **17.** 173 000 **18.** 29 800 **19.** 6 750 000 **20.** 1070

Write these correct to 3 significant figures:

21. 8047 **22.** 9143 **23.** 8765 **24.** 9998 **25.** 3 762 000
26. 980 930 **27.** 6 975 000 **28.** 10 950 **29.** 7 896 000 **30.** 99 999

Write these correct to 1 significant figure:

31. 4.6 **32.** 0.83 **33.** 0.55 **34.** 0.083 **35.** 0.0076
36. 0.0306 **37.** 0.704 **38.** 0.98 **39.** 0.0098 **40.** 0.000 508

Write these correct to 2 significant figures:

41. 0.79 **42.** 0.486 **43.** 0.315 **44.** 0.0417 **45.** 0.002 34
46. 0.1073 **47.** 0.0632 **48.** 0.0507 **49.** 0.0974 **50.** 0.090 99

Write these correct to 3 significant figures:

51. 8.439 **52.** 0.9437 **53.** 0.8255 **54.** 0.010 37 **55.** 0.100 43
56. 0.170 98 **57.** 0.4299 **58.** 0.050 55 **59.** 0.007 539 **60.** 0.9999

Write each number correct to 1, 2 and 3 significant figures:

61. 7394 **62.** 435 600 **63.** 11.73 **64.** 276.4 **65.** 0.4073
66. 29.06 **67.** 40.404 **68.** 8.375 **69.** 0.9099 **70.** 0.037 55

Exercise 21.4 *Estimates*

Find an approximate value for each of the following correct to 1 significant figure:

1. $430 + 560$ **2.** $60\,000 + 120$ **3.** $1200 - 801$
4. $800 - 0.08$ **5.** $0.7 + 1.4 - 0.8$ **6.** 8.3×4.1
7. 4.9×1.8 **8.** 3.06×2.1 **9.** $(0.3)^3$
10. 18.39×0.5 **11.** 63.7×9.8 **12.** 42.3×41
13. $3.7 \times 2.1 \times 8.95$ **14.** $700 \times 1.5 \times 0.48$ **15.** $27.4 \times 0.4 \times 8$
16. $80.7 \div 8.9$ **17.** $372 \div 21$ **18.** $83 \div 1.9$
19. $47 \div 2.01$ **20.** $3.85 \div 4.1$ **21.** $127 \div 5.5$
22. $31.4 \div 0.5$ **23.** $8.6 \div 0.4$ **24.** $38 \div 0.2$ **25.** $4.3 \div 0.08$

Find an approximate value for each of the following correct to 2 significant figures:

26. $238 + 3.62 + 49.3$ **27.** $2.19 + 0.87 + 0.006$ **28.** $402.6 - 198.7$
29. $8.8 - 0.087$ **30.** $1.4 + 600 - 18.7$ **31.** 36.7×21
32. 28.7×0.5 **33.** 7.5×4.06 **34.** 180×29.6
35. $(0.5)^3 \times 3$ **36.** $7.5 \div 2.95$ **37.** $90.7 \div 20.05$
38. $1.71 \div 0.5$ **39.** $\sqrt{1000}$ **40.** $\sqrt{3000}$

Exercise 21.5 *Harder estimates*

Find an approximate value for each of the following correct to 1 significant figure:

1. $\dfrac{4.2 + 3.9}{1.4 + 1.6}$ **2.** $\dfrac{18.7 - 9.4}{18.4 + 9.7}$ **3.** $\dfrac{14.1 + 141}{0.7 - 0.2}$

4. $\dfrac{185 + 1.9}{190 - 0.07}$ **5.** $\dfrac{4.7 \times 2.1}{3}$ **6.** $\dfrac{24}{1.5 \times 3.8}$

7. $\dfrac{3.07 \times 2.4}{(0.9)^2}$ **8.** $\dfrac{14.7 \times 8.9}{2.3 \times 3.1}$ **9.** $\dfrac{307 \times 0.5}{12.3 \times 13}$

10. $\dfrac{0.37 \times 6}{(2.1)^2}$ **11.** $\dfrac{238 \times 19.6}{43.6}$ **12.** $\dfrac{73.4 \times 27}{0.5 \times 4.3}$

13. $\dfrac{60.7 \times 83}{33.3}$ **14.** $\dfrac{41.7 \times 59.2}{83.7 \times 30.4}$ **15.** $\dfrac{426 \times 0.101}{161}$

Exercise 21.6 *Conversion from standard form*

Write these as ordinary numbers (without powers of 10):

1.	1.7×10^2	**2.**	3.95×10^3	**3.**	4.7×10^3	**4.**	8×10^4
5.	6.9×10^6	**6.**	4.3×10^{-1}	**7.**	2.95×10^{-2}	**8.**	5.7×10^{-2}
9.	8.3×10^{-3}	**10.**	4×10^{-4}	**11.**	8.4×10^3	**12.**	8.4×10^{-3}
13.	4.72×10^5	**14.**	4.72×10^{-5}	**15.**	7.03×10^4	**16.**	7.03×10^{-4}
17.	2×10^2	**18.**	2×10^{-2}	**19.**	6×10^0	**20.**	6×10^{-6}

Exercise 21.7 *Conversion to standard form*

Write these in standard form:

1. 460 **2.** 21 000 **3.** 8000 **4.** 47 500 **5.** Six million
6. 0.7 **7.** 0.495 **8.** 0.0092 **9.** 0.000 06 **10.** 0.019

11.	43 700	**12.**	0.17	**13.**	0.007
14.	19 000 000	**15.**	Twenty eight thousand	**16.**	0.6
17.	Three hundredths	**18.**	27 600	**19.**	0.8
20.	8	**21.**	27.6×10^3	**22.**	192×10^4
23.	18×10^{-2}	**24.**	0.03×10^{-4}	**25.**	425×10^{-1}

Exercise 21.8 *Calculations with standard form*

Work out the following, giving your answer in standard form:

1.	$4 \times 10^4 \times 2 \times 10^2$	**2.**	$1.6 \times 10^3 \times 4 \times 10^5$
3.	$5 \times 10^1 \times 6 \times 10^2$	**4.**	$3.5 \times 10^3 \times 3 \times 10^4$
5.	$1.25 \times 10^2 \times 7 \times 10^1$	**6.**	$3 \times 10^{-4} \times 2 \times 10^6$
7.	$4 \times 10^{-1} \times 2 \times 10^{-3}$	**8.**	$1.2 \times 10^4 \times 3 \times 10^{-1}$
9.	$4.6 \times 10^{-2} \times 3 \times 10^1$	**10.**	$17.3 \times 10^3 \times 7 \times 10^{-4}$
11.	$8 \times 10^4 \div (2 \times 10^3)$	**12.**	$8 \times 10^3 \div (2 \times 10^5)$
13.	$8 \times 10^3 \div (4 \times 10^{-2})$	**14.**	$8 \times 10^{-3} \div (4 \times 10^2)$
15.	$9.6 \times 10^4 \div (3 \times 10^3)$	**16.**	$7.4 \times 10^2 \div (2 \times 10^{-3})$
17.	$1.44 \times 10^{-2} \div (1.2 \times 10^2)$	**18.**	$8.1 \times 10^4 \div (3 \times 10^3)$
19.	$2.45 \times 10^{-1} \div (5 \times 10^{-4})$	**20.**	$1.38 \times 10^{-3} \div (6 \times 10^1)$
21.	$\dfrac{6 \times 10^3 \times 3 \times 10^4}{2 \times 10^2}$	**22.**	$\dfrac{1.8 \times 10^2 \times 9 \times 10^1}{6 \times 10^4}$
23.	$\dfrac{1.4 \times 10^{-4} \times 7.2 \times 10^5}{6 \times 10^2}$	**24.**	$\dfrac{2.7 \times 10^6}{9 \times 10^3 \times 2 \times 10^2}$
25.	$\dfrac{5.4 \times 10^3}{3 \times 10^2 \times 9 \times 10^{-4}}$	**26.**	$\dfrac{3.75 \times 10^{-4}}{3 \times 10^{-3} \times 5 \times 10^2}$
27.	$\dfrac{270 \times 0.14}{90}$	**28.**	$\dfrac{81}{0.09 \times 30}$
29.	$\dfrac{64 \times 0.36}{0.8 \times 900}$	**30.**	$\dfrac{14.4 \times 82}{0.06 \times 0.41}$

22

Arithmetic in bases other than ten

Exercise 22.1 *Addition*

Do these additions in the given base:

1. $110 + 101$ (base 2)
2. $101 + 11$ (base 2)
3. $1011 + 110$ (base 2)
4. $212 + 21$ (base 3)
5. $1012 + 211$ (base 3)
6. $312 + 223$ (base 4)
7. $3210 + 333$ (base 4)
8. $342 + 214$ (base 5)
9. $3104 + 423$ (base 5)
10. $413 + 454$ (base 6)
11. $3215 + 542$ (base 6)
12. $413 + 564$ (base 7)
13. $6214 + 524$ (base 7)
14. $732 + 664$ (base 8)
15. $361 + 746$ (base 8)
16. $841 + 637$ (base 9)
17. $4217 + 837$ (base 9)
18. $1011 + 100 + 1011$ (base 2)
19. $1011 + 1101 + 1101$ (base 2)
20. $212 + 122 + 221$ (base 3)
21. $312 + 23 + 132$ (base 4)
22. $244 + 1234 + 213$ (base 5)
23. $215 + 2532 + 435$ (base 6)
24. $1246 + 353 + 624$ (base 7)
25. $376 + 542 + 77$ (base 8)
26. $837 + 214 + 8621$ (base 9)
27. $214 + 145 + 32$ (base 6)
28. $1021 + 212 + 122$ (base 3)
29. $314 + 424 + 133$ (base 5)
30. $747 + 707 + 727$ (base 8)
31. $2313 + 323 + 301$ (base 4)
32. $243 + 146 + 260$ (base 7)
33. $1001 + 110 + 1011 + 11$ (base 2)
34. $1110 + 111 + 1001 + 1101$ (base 2)
35. $1110 + 11 + 1010 + 101$ (base 2)

Exercise 22.2 *Subtraction*

Do these subtractions in the given base:

1. $1011 - 101$ (base 2)
2. $215 - 36$ (base 8)
3. $214 - 35$ (base 6)
4. $1001 - 111$ (base 2)
5. $200 - 121$ (base 3)
6. $332 - 123$ (base 4)
7. $831 - 658$ (base 9)
8. $234 - 65$ (base 7)
9. $423 - 214$ (base 5)
10. $600 - 142$ (base 8)
11. $10\,010 - 1011$ (base 2)
12. $2100 - 122$ (base 3)
13. $3401 - 423$ (base 5)
14. $7000 - 1342$ (base 8)
15. $100\,001 - 11\,010$ (base 2)
16. $2731 - 685$ (base 9)
17. $2300 - 312$ (base 4)
18. $1320 - 354$ (base 6)
19. $1432 - 564$ (base 7)
20. $100\,000 - 1101$ (base 2)
21. $1000 - 121$ (base 3)
22. $1000 - 121$ (base 4)
23. $100\,110 - 1001$ (base 2)
24. $4200 - 437$ (base 8)

Exercise 22.3 *Multiplication of single digits*

Multiply in base 3:

1. 2×2

Multiply in base 4:

2. 2×2
3. 2×3
4. 3×3

Multiply in base 5:

5. 2×3 **6.** 2×4 **7.** 3×3 **8.** 3×4 **9.** 4×4

Multiply in base 6:

10. 2×3 **11.** 2×4 **12.** 2×5 **13.** 3×3
14. 3×4 **15.** 3×5 **16.** 4×4 **17.** 4×5 **18.** 5×5

Multiply in base 7:

19. 2×4 **20.** 2×5 **21.** 2×6 **22.** 3×3
23. 3×4 **24.** 3×5 **25.** 3×6 **26.** 4×4
27. 4×5 **28.** 4×6 **29.** 5×5 **30.** 5×6 **31.** 6×6

Multiply in base 8:

32. 2×4 **33.** 2×5 **34.** 2×6 **35.** 2×7 **36.** 3×3
37. 3×4 **38.** 3×5 **39.** 3×6 **40.** 3×7 **41.** 4×4
42. 4×5 **43.** 4×6 **44.** 4×7 **45.** 5×5 **46.** 5×6
47. 5×7 **48.** 6×6 **49.** 6×7 **50.** 7×7

Multiply in base 9:

51. 2×5 **52.** 2×6 **53.** 2×7 **54.** 2×8 **55.** 3×3
56. 3×4 **57.** 3×5 **58.** 3×6 **59.** 3×7 **60.** 3×8
61. 4×4 **62.** 4×5 **63.** 4×6 **64.** 4×7 **65.** 4×8
66. 5×5 **67.** 5×6 **68.** 5×7 **69.** 5×8 **70.** 6×6
71. 6×7 **72.** 6×8 **73.** 7×7 **74.** 7×8 **75.** 8×8

Exercise 22.4 *Short multiplication*

Do these multiplications in the given base:

1. 121×2 (base 3) **2.** 234×4 (base 5) **3.** 143×6 (base 8)
4. 123×3 (base 4) **5.** 627×5 (base 9) **6.** 324×5 (base 7)
7. 215×3 (base 6) **8.** 174×2 (base 8) **9.** 212×2 (base 3)
10. 345×4 (base 6) **11.** 3426×4 (base 9) **12.** 4315×5 (base 8)
13. 1321×3 (base 4) **14.** 4123×4 (base 5) **15.** 3225×3 (base 6)
16. 2473×6 (base 8) **17.** 5721×8 (base 9) **18.** 1022×2 (base 3)
19. 1234×3 (base 5) **20.** 4352×5 (base 7)

Exercise 22.5 *Long multiplication*

Multiply these in binary:

1. 1011×11 **2.** 1101×11 **3.** $10\,111 \times 11$ **4.** $11\,001 \times 101$
5. $11\,011 \times 101$ **6.** $10\,111 \times 110$ **7.** $10\,011 \times 1011$ **8.** $11\,011 \times 1101$
9. $10\,001 \times 1110$ **10.** $11\,101 \times 1001$

Multiply these in the base indicated:

11. 436×23 (base 8) **12.** 132×42 (base 5) **13.** 352×25 (base 6)
14. 615×34 (base 7) **15.** 102×21 (base 3) **16.** 374×37 (base 9)

17. 232 × 32 (base 4)
18. 1243 × 214 (base 5)
19. 3233 × 231 (base 4)
20. 1221 × 220 (base 3)

Exercise 22.6 *Short division*

Do these divisions in the given base:

1. 2602 ÷ 5 (base 8)
2. 723 ÷ 4 (base 9)
3. 2211 ÷ 3 (base 4)
4. 1223 ÷ 3 (base 6)
5. 2432 ÷ 5 (base 7)
6. 1234 ÷ 2 (base 5)
7. 1210 ÷ 2 (base 3)
8. 2112 ÷ 3 (base 4)
9. 2332 ÷ 6 (base 8)
10. 1122 ÷ 3 (base 5)
11. 4057 ÷ 5 (base 8)
12. 1232 ÷ 4 (base 5)
13. 3203 ÷ 7 (base 9)
14. 2110 ÷ 2 (base 3)
15. 2154 ÷ 6 (base 7)
16. 10 212 ÷ 2 (base 4)
17. 10 212 ÷ 2 (base 3)
18. 10 131 ÷ 3 (base 5)
19. 23 532 ÷ 6 (base 8)
20. 23 532 ÷ 5 (base 6)

Exercise 22.7 *Long division in binary*

Divide these in binary:

1. $1100_2 \div 10_2$
2. $1001_2 \div 11_2$
3. $10\,010_2 \div 11_2$
4. $11\,000_2 \div 110_2$
5. $10\,100_2 \div 100_2$
6. $101\,000_2 \div 101_2$
7. $100\,100_2 \div 100_2$
8. $110\,010_2 \div 101_2$
9. $100\,011_2 \div 111_2$
10. $100\,100_2 \div 110_2$
11. $110\,001_2 \div 111_2$
12. $101\,010_2 \div 1110_2$
13. $110\,111_2 \div 1011_2$
14. $111\,000_2 \div 1000_2$
15. $110\,110_2 \div 1001_2$
16. $1\,010\,100_2 \div 1100_2$
17. $1\,000\,110_2 \div 1010_2$
18. $1\,101\,001_2 \div 1111_2$
19. $10\,010\,000_2 \div 1100_2$
20. $1\,111\,110_2 \div 1110_2$

23

Further work with sets

Exercise 23.1 *Introduction to Venn diagrams involving three sets*

1. Copy the figure 5 times. Shade in each of these on a different drawing:

   (a) $A \cap B$ (b) $A \cap B \cap C'$
   (c) $A \cap B' \cap C'$ (d) $A \cup B$
   (e) $A \cup B'$

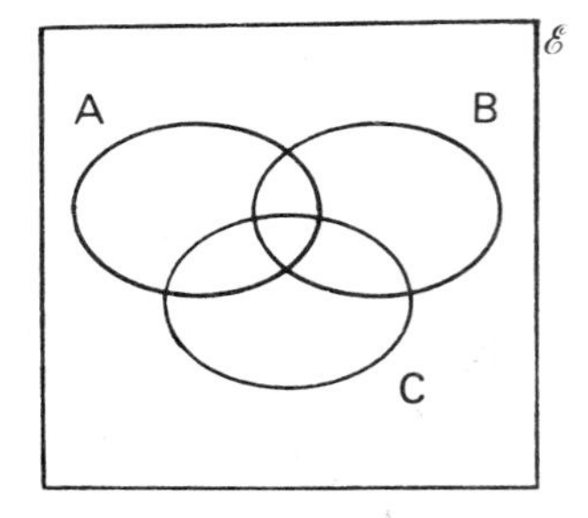

2. Describe each of the shaded regions in set language:

   (a)

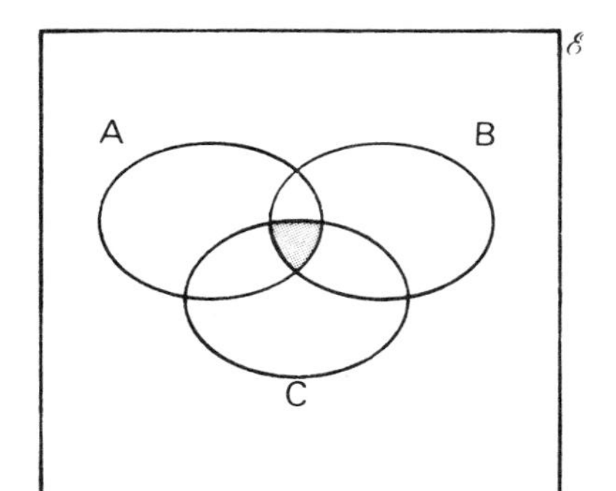

   (b)

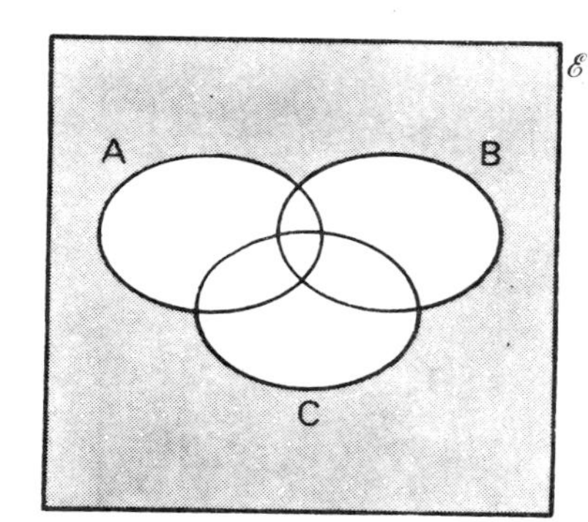

3. Write down the number of elements in each of these:

   (a) Q (b) $P \cap Q$
   (c) $P \cap Q' \cap R'$ (d) $P \cup R$
   (e) $R \cap P'$ (f) $(P \cup Q \cup R)'$

   Write down the members of:

   (g) R (h) $Q \cap R$
   (i) $Q \cap P'$ (j) $R \cup Q$
   (k) $(P \cap R) \cup Q$ (l) $(P \cup R) \cap Q$

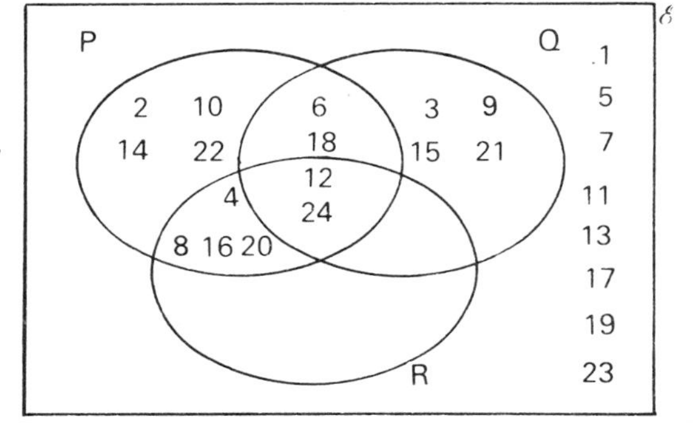

4. Draw Venn diagrams to illustrate the following regions:

   (a) $(A \cup B) \cap C$ (b) $(A \cup B)' \cap C$ (c) $(A \cap B)' \cap C$
   (d) $(A \cup B) \cap C'$ (e) $(A \cap B) \cup (A \cap C)$ (f) $(A \cup B) \cap (A \cup C)$

5. Show that the following statements are true:
   (a) $A \cup (B \cup C) = (A \cup B) \cup C$
   (b) $A \cup (B \cap C) = (A \cup B) \cap (A \cup C)$
   (c) $(A \cap B)' = A' \cup B'$

Exercise 23.2 *Statements, Venn diagrams and set notation*

1. $\mathscr{E}$ = {children at Holiday Hall}
   C = {those who like cereal}
   P = {those who like porridge}
   Describe the likes and dislikes of:
   (a) $C \cap P$ (b) $C \cup P$ (c) $C \cap P'$ (d) $(C \cup P)'$
2. F = {those who bite their finger nails}
   T = {those who bite their toe nails}
   What do the following do?:
   (a) $F \cap T$ (b) $(F \cap T)'$ (c) $F' \cap T$ (d) F'
3. S = {saloon cars}
   B = {blue cars}
   A = {cars made by Autosmall}
   What types of cars are represented by:
   (a) $S \cap B$ (b) $S \cup B$ (c) $S \cap A \cap B$
   (d) $(S \cap A) \cup B$ (e) $B \cap S'$ (f) $(A' \cap S) \cap B$?
   (g) If $B \cap A = \emptyset$, what deductions can you make?
4. F = {those who learn French}
   G = {those who learn German}
   R = {those who learn Russian}
   S = {those who learn Spanish}
   What can be said about the language choices of these groups?
   (a) $F \cap G$ (b) $F \cap G \cap R \cap S$
   (c) $R \cup S$ (d) $(G \cup R) \cap F$
   (e) $G \cap R'$ (f) $F \cap G' \cap S'$
   (g) $(R \cap S)'$ (h) $(R \cup S)'$
5. $\mathscr{E}$ = {boys at St. Gilbert's School}
   C = {boys who play in a school cricket team}
   F = {boys who play in a school football team}
   R = {boys who play in a school rugby team}

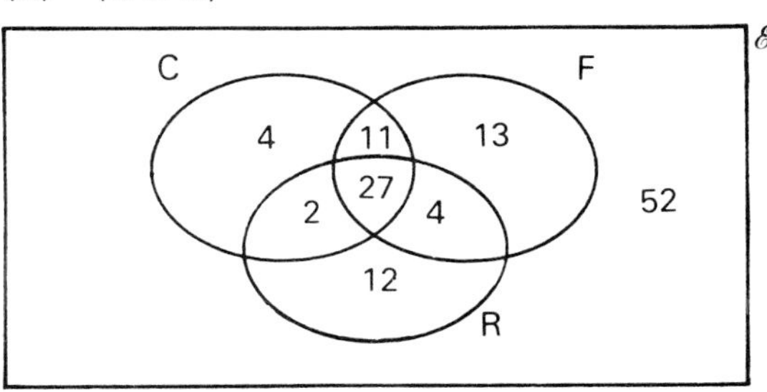

   (a) How many boys are there at St. Gilbert's?
   (b) How many boys are there in the rugby teams?
   (c) How many cricket XIs are there?
   (d) How many are only in a football team?
   (e) How many play all three games for the school?
   (f) How many play cricket and football for the school?
   (g) How many play rugby and football only for the school?
   (h) How many play only two games for the school?

6. 10 pupils in a class play the piano, 6 the trumpet and 12 the recorder. Of these 2 play all three instruments, 1 plays the trumpet only and 3 the piano only while 1 plays trumpet and recorder only.
How many play only the recorder?

7. Three quarters of the pupils in a school of 100 do carpentry, shooting or dancing as an 'extra'.
62 take carpentry. 45 shoot. 36 do both carpentry and shooting, of whom 6 dance as well. 24 do carpentry only and 8 only visit the shooting range.
How many (**a**) dance and shoot (**b**) dance and do carpentry only (**c**) only dance?

8. All the pupils at a day school travel either by bus or train. $\frac{7}{8}$ catch a bus while $\frac{1}{4}$ catch a train. If there 240 pupils altogether, how many use both means of transport?

9. Some children had a choice of fruit. There were 15 apples, 10 bananas and 17 peaches. Nobody was allowed all three. 6 took an apple and a banana, 4 a banana and a peach and 3 chose an apple and a peach. 6 children ate nothing, and the others took 1 fruit each. If all the fruit was eaten, how many children were present?

10. D = {pupils who own a dog}
C = {pupils who own a cat}
P = {pupils who own a pony}
The following facts are known:
$n(D \cap C \cap P) = 2$
$n(D \cap P) = 3$
$n(C \cap P) = 5$
$n(D \cap C' \cap P') = 6$
$n(C \cap D' \cap P') = 8$
If $n(D \cup C) = 24$ and $n(D \cup C \cup P) = 26$, find $n(D \cap C \cap P')$ and $n(P)$.

24

Harder areas and volumes

Exercise 24.1 *Parallelogram, triangle and trapezium*

Questions 1–25 refer to the parallelogram ABCD on the right.

Find the area of ABCD when:

1. AB = 8 cm, EF = 5 cm
2. AB = 12 cm, EF = 3 cm
3. AB = 2 m, EF = 1.6 m
4. BC = 1.5 cm, GH = 3 cm
5. BC = 1.8 m, GH = 0.5 m
6. BC = 3.5 cm, GH = 3.2 cm
7. CD = 4 m, EF = 1.2 m
8. CD = 15 cm, EF = 18 cm
9. DA = 14 cm, GH = 1.5 cm
10. DA = 3.6 m, GH = 2.5 m
11. AB = 5.4 m, EF = 60 cm (answer in m^2)
12. AB = 750 m, EF = 600 m (answer in km^2)
13. If AB = 8 cm and the area is 24 cm^2, find EF.
14. If AB = 3 m and the area is 10.5 m^2, find EF.
15. If AB = 6 cm and the area is 14.4 cm^2, find EF.
16. If EF = 2 m and the area is 8 m^2, find AB.
17. If EF = 9 cm and the area is 72 cm^2, find AB.
18. If EF = 1.6 m and the area is 12.8 m^2, find CD.
19. If GH = 5 cm and the area is 35 cm^2, find AD.
20. If BC = 8 cm and the area is 88 cm^2, find GH.
21. If AB = 4 cm, EF = 6 cm, AD = 8 cm, find the area. Hence find GH.
22. If BC = 10 cm, GH = 4 cm, EF = 5 cm, find the area. Hence find AB.
23. If AD = 15 cm, GH = 8 cm, EF = 10 cm, find CD.
24. If CD = 6 cm, AD = 8 cm, EF = 4 cm, find GH.
25. If EF = 5 cm, GH = 6 cm, CB = 10 cm, find AB.

Questions 26–42 refer to the triangle ABC on the right.

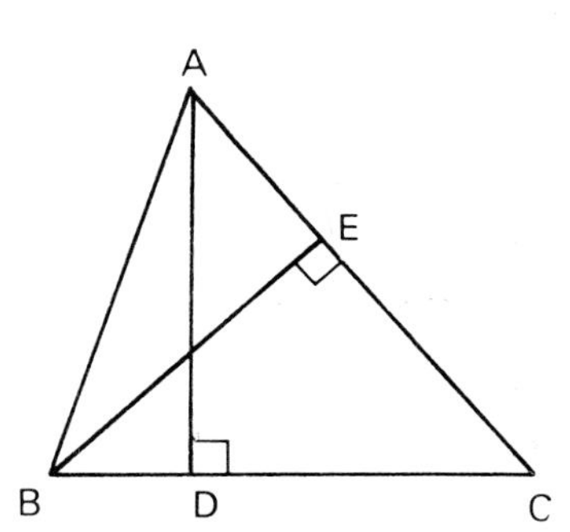

26. If BC = 6 cm, AD = 5 cm, find the area of the triangle.
27. If BC = 3 m, AD = 6 m, find the area of the triangle.
28. If BC = 8 cm, AD = 9 cm, find the area of the triangle.

29. If AC = 4 cm, BE = 3 cm, find the area of the triangle.
30. If AC = 15 cm, BE = 8 cm, find the area of the triangle.
31. If BC = 6.4 cm, AD = 6 cm, find the area of the triangle.
32. If BC = 4.8 m, AD = 3 m, find the area of the triangle.
33. If BC = 5.2 m, AD = 3.5 m, find the area of the triangle.
34. If AD = 6 cm and the area is 9 cm^2, find BC.
35. If AD = 12 cm and the area is 42 cm^2, find BC.
36. If AD = 24 cm and the area is 150 cm^2, find BC
37. If AD = 15 cm and the area is 120 cm^2, find BC.
38. If BC = 8 cm and the area is 40 cm^2, find AD.
39. If BC = 25 cm and the area is 300 cm^2, find AD.
40. If AD = 6 cm, BC = 12 cm, AC = 9 cm, find the area. Hence find BE.
41. If BC = 8 cm, AD = 6 cm, BE = 4 cm, find AC.
42. If BC = 8 cm, AC = 12 cm, BE = 6 cm, find AD.

Find the areas of the following:

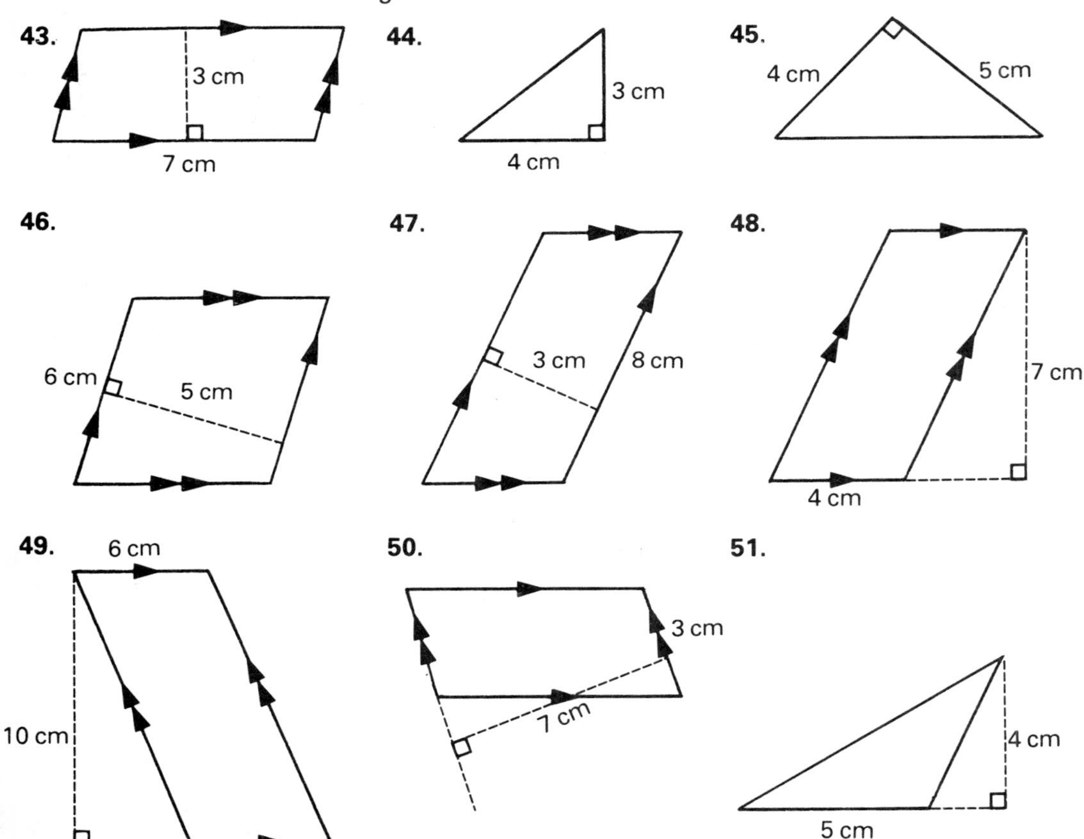

52. A triangle has base 15 cm and height 14 cm. Find its area.
53. A parallelogram has base 22 cm and height 8 cm. Find its area.
54. Find the area of a triangle with base 4.2 m and height 1.6 m.
55. A parallelogram has area 56 cm^2 and base 14 cm. Find its height.
56. A triangle has height 12 cm and area 84 cm^2. Find the base.
57. Find the height of a triangle with area 42 cm^2 and base 12 cm.

Find the areas of the following trapezia:

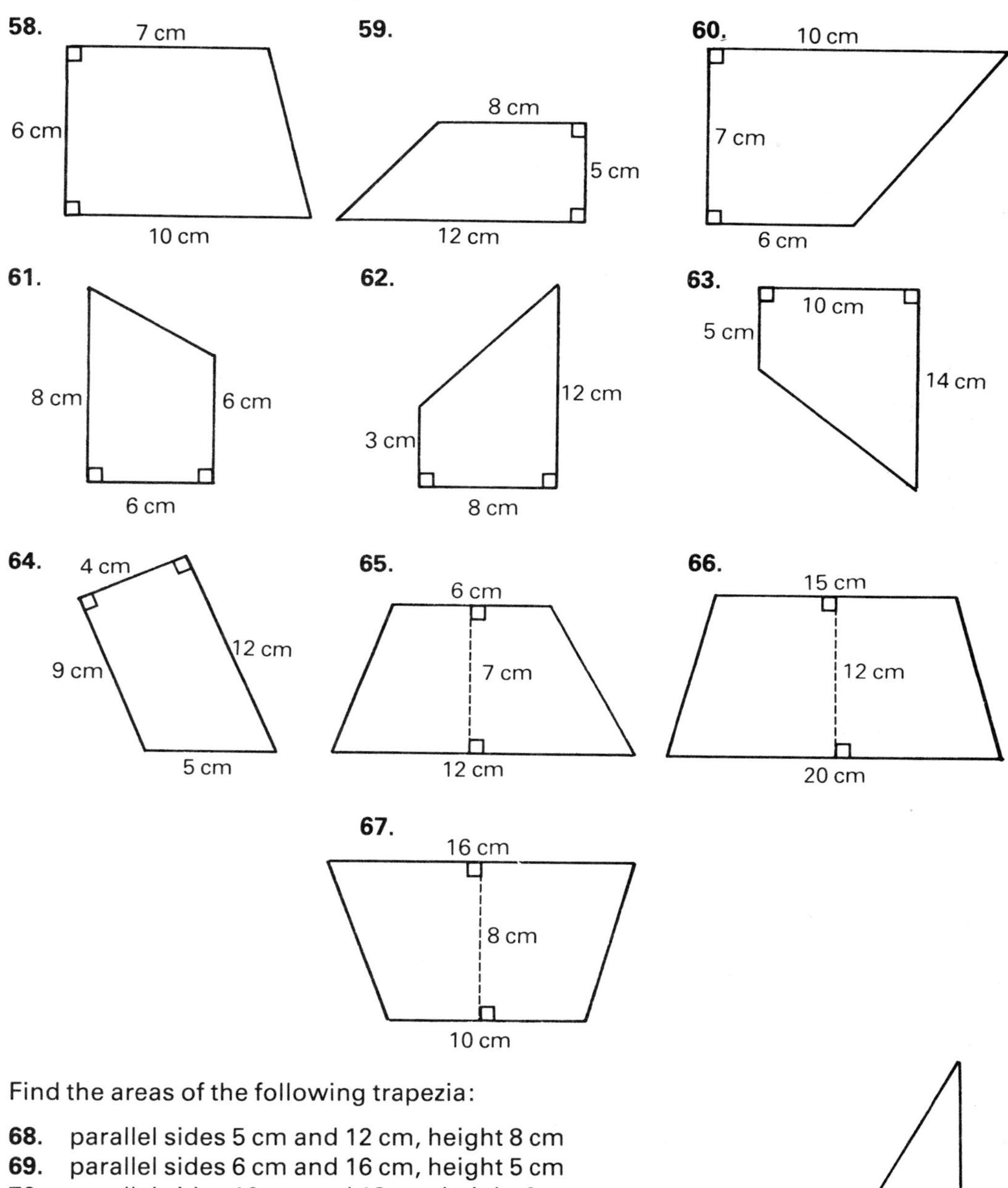

Find the areas of the following trapezia:

68. parallel sides 5 cm and 12 cm, height 8 cm
69. parallel sides 6 cm and 16 cm, height 5 cm
70. parallel sides 10 cm and 18 cm, height 6 cm
71. height 4 cm, parallel sides 8 cm and 6 cm

72. The diagram on the right shows a sail. Find its area.

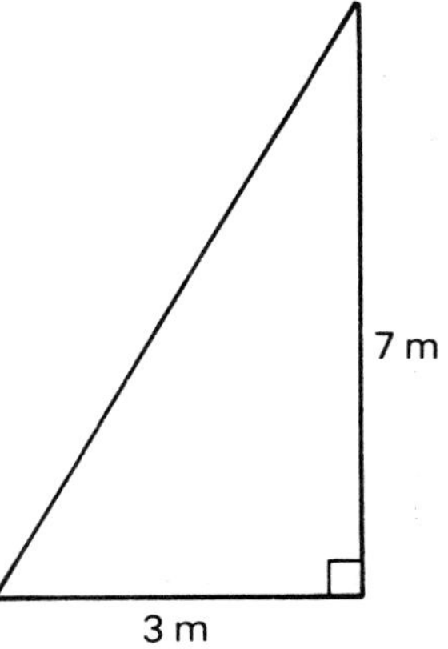

73. The diagram shows part of the roof of a house. Find its area.

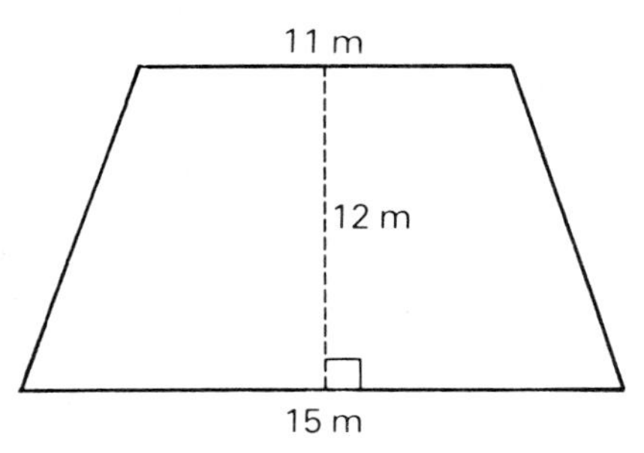

74. A field is a trapezium with two parallel sides of length 220 m and 340 m. One of the other sides is perpendicular to these two sides and is 250 m long. Find the area of the field in km^2.

Exercise 24.2 *Composite bodies – surface area*

Find the areas of these figures:

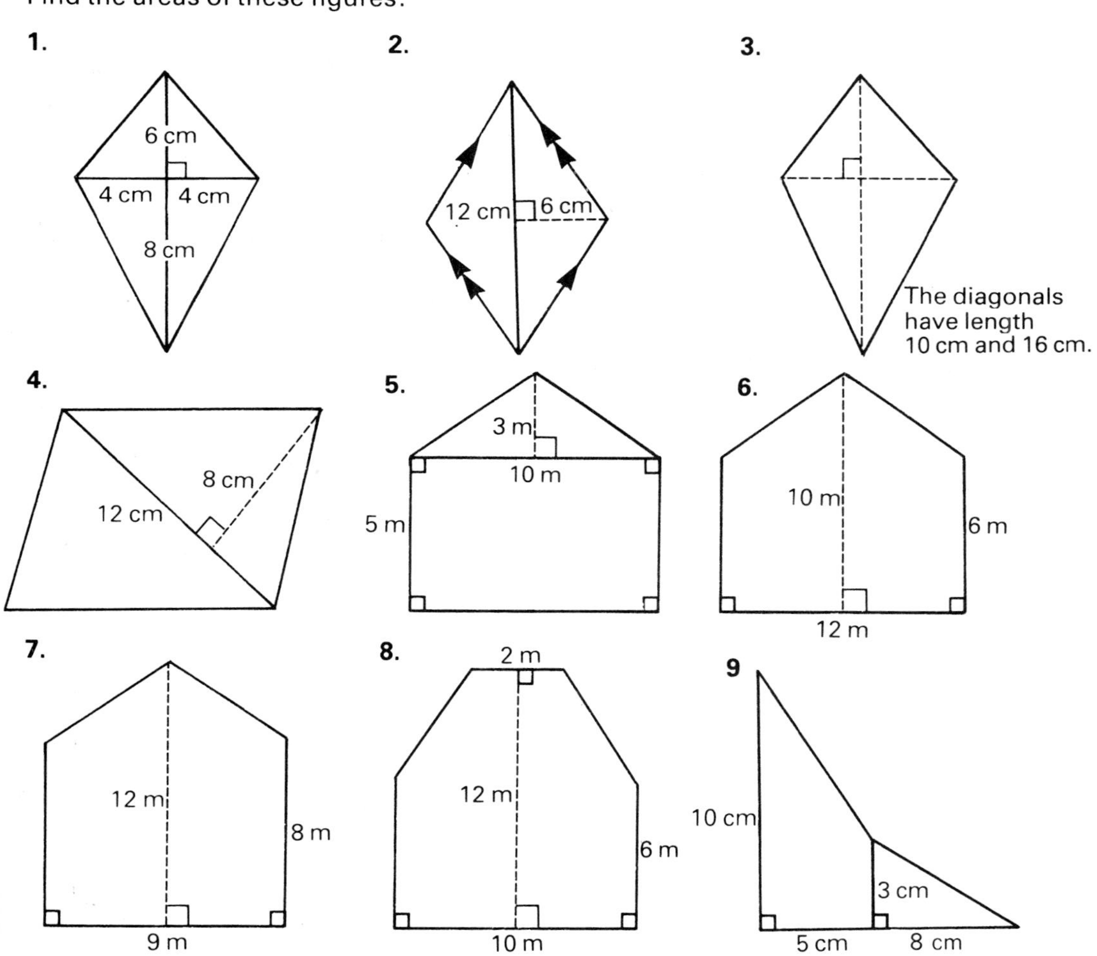

10. A living room is 8 m long and 3 m wide. The walls are 2.5 m high. Find the total area of the four walls and ceiling.
11. A kitchen is 5 m long and 3 m wide. The walls are 2.2 m high. Find the total area of the four walls and ceiling.
12. An upstairs landing is 15 m long, 2 m wide and 3 m high. In its walls are 4 doors, each 1.2 m wide and 2 m high, and a window 1.6 m wide and 1.5 m high. Find the area of the remainder of the walls.
13. Find the surface area of a tool box measuring 50 cm by 30 cm by 25 cm.
14. A shoe box without a lid is 30 cm long, 15 cm wide and 12 cm high. Find the area of cardboard used in making it.
15. A wooden pencil case measures 25 cm by 8 cm by 5 cm. Find its surface area.
16. Find the surface area of a cardboard box (with lid), measuring 60 cm by 40 cm by 50 cm. Answer in m^2.
17. Find the surface area of a cereal packet measuring 30 cm by 20 cm by 8 cm.
18. The figure on the right shows the front of an A-frame tent 2.5 m long, 1.8 m wide and 1.2 m high. The sloping edges are each 1.5 m long. Find the area of canvas. (Do not include the groundsheet.)

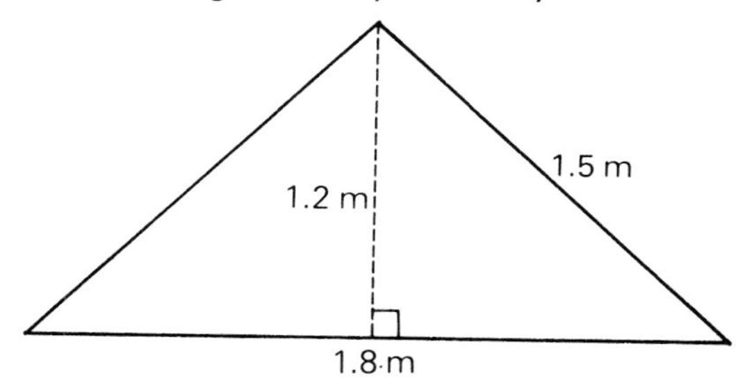

19. The front of another tent is shown on the right. This tent is 3 m long. Find its surface area.

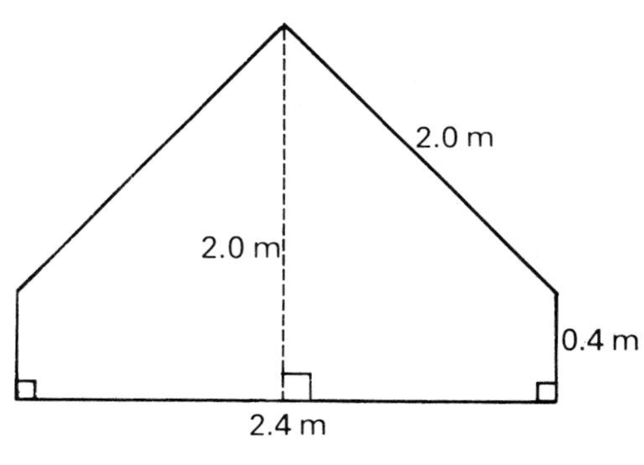

20. The figure on the right shows the front of a garden shed 5 m long. The roof measures 5.6 m by 3.5 m. Find the total area of the floor, the four walls and the roof.

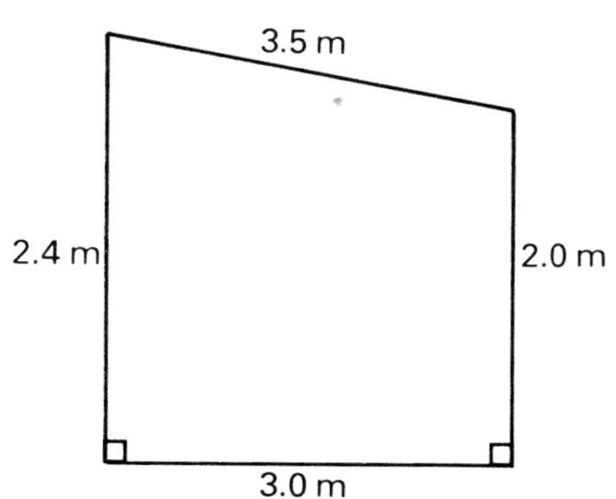

Exercise 24.3 *Volumes of cuboids*

Find the volumes of cuboids with the given sides:

	Length	Width	Height
1.	3 m	2 m	2 m
2.	4 m	3 m	2 m
3.	10 cm	8 cm	4 cm
4.	8 cm	5 cm	3 cm
5.	6 cm	4 cm	4 cm
6.	4 m	2.4 m	1.5 m
7.	5 m	4.2 m	3 m
8.	Square base of side 8 cm		10 cm

9. A cuboid has volume 60 cm^3, length 5 cm, width 4 cm. Find the height.
10. A cuboid has height 2 cm, width 5 cm, volume 80 cm^3. Find the length.
11. A cuboid has length 10 cm, height 4 cm, volume 240 cm^3. Find the width.
12. A cuboid has volume 150 cm^3 and height 6 cm. Find the base area.

In Questions 13–19 V is the volume of a cuboid, A its base area, l its length, w its width and h its height.

13. $l = 6$ cm, $w = 5$ cm, $h = 4$ cm. Find V.
14. $l = 8$ cm, $w = 6$ cm, $V = 96$ cm^3. Find h.
15. $V = 240$ m^3, $w = 8$ m, $h = 3$ m. Find l.
16. $h = 1.2$ m, $V = 6$ m^3, $w = 2$ m. Find l.
17. $A = 4.5$ m^2, $h = 3$ m. Find V.
18. $V = 20$ cm^3, A = 8 cm^2. Find h.
19. $h = 3.6$ cm, $V = 144$ cm^3. Find A.

(Long multiplication) Find the volume of each of these:

20. A room 4.5 m long, 3.2 m wide and 2.0 m high.
21. A box 28 cm long, 25 cm wide and 12 cm high.
22. A book 22 cm tall, 15 cm wide and 8 cm thick.
23. A sideboard 2.4 m wide, 1.8 m high and 0.75 m deep.
24. A wardrobe 2.1 m tall, 1.8 m wide and 0.6 m deep.
25. A chest freezer 2.4 m long, 0.9 m wide and 1.25 m deep.
26. A telephone kiosk 0.8 m square and 2.4 m high.
27. A letter box 0.5 m square and 1.2 m high.

(Mixed units) Find the volume of each of these.

28. A box of cough sweets 12 cm long, 6 cm wide and 8 mm thick. (Answer in cm^3.)
29. A paving stone 0.8 m square and 2.5 cm thick. (Answer in cm^3.)
30. A laminated table top measuring 1.2 m $\times$ 0.8 m $\times$ 6 mm. (Answer in cm^3.)
31. A sheet of metal 2 m square and 0.2 mm thick. (Answer in cm^3.)
32. A telephone directory measuring 32 cm by 20 cm by 15 mm. (Answer in cm^3.)
33. A wall 60 m long, 4 m high and 20 cm thick. (Answer in m^3.)

34. The tarmac on a runway 6 km long, 20 m wide and 3 cm thick. (Answer in m^3.)
35. A piece of metal measuring 5 mm × 1 mm × 2 m. (Answer in cm^3.)

(Composite bodies) Find the volume of each of these:

36. The bracket in Fig. 1. It is 4 cm long.
37. The clamp in Fig. 2. It is 2 cm long.
38. The girder in Fig. 3. It is 4 m long.
39. The step-block in Fig. 4. It is 1.5 m wide.
40. The girder in Fig. 5. It is 3 m long.

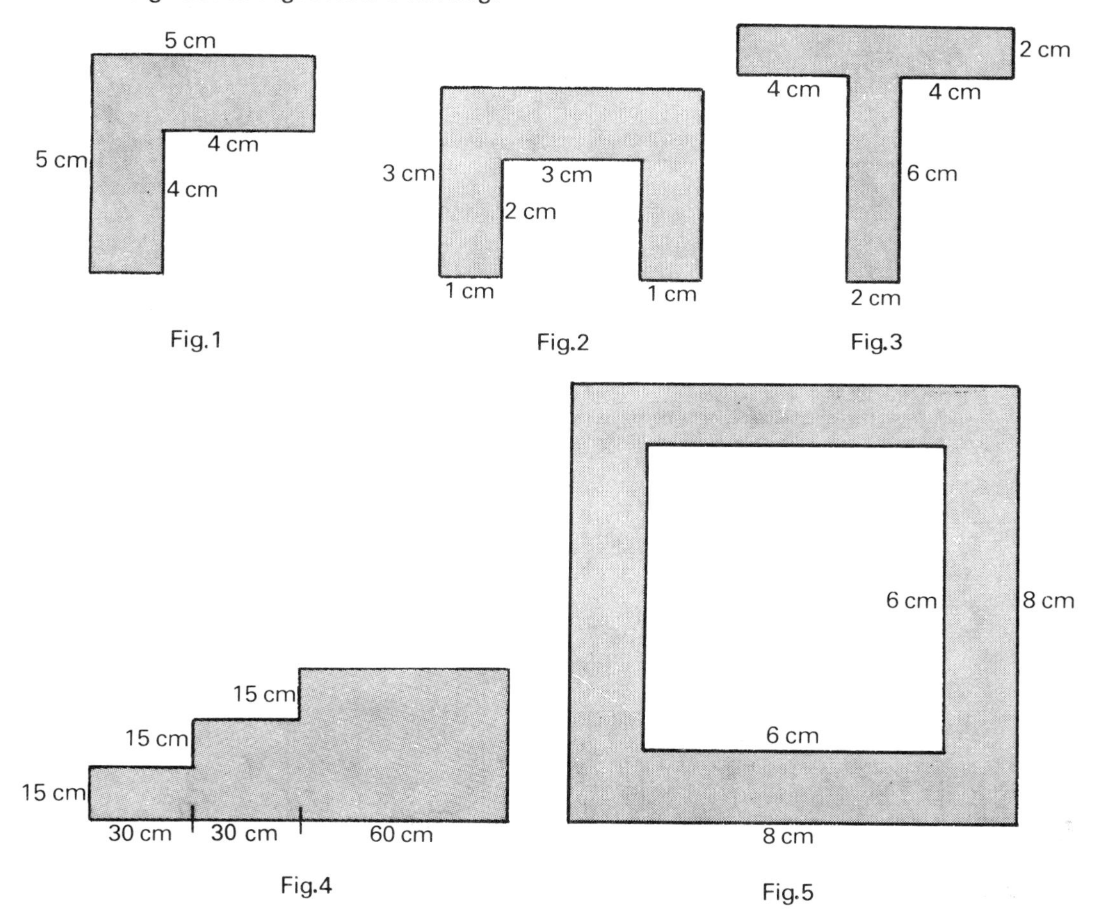

41. The metal in a metal box with external measurements 10 cm × 8 cm × 4 cm and internal measurements 9 cm by 7 cm by 3 cm.
42. The wood in a wooden box with external measurements 60 cm × 45 cm × 25 cm and internal measurements 55 cm × 40 cm × 20 cm.

43. (Prisms) Find the volumes of the tents and shed in Exercise 24.2, Questions 18, 19 and 20.

Exercise 24.4 *Miscellaneous problems – mass, capacity and cost*

1. A can of weed-killer costs £12 and will cover 200 m^2 of lawn. Find the cost of covering a lawn measuring 20 m by 30 m.
2. A tin of paint costs £10 and will cover 10 m^2 of wall. Find the cost of painting a wall 4 m long and 2 m high.
3. A barrel of tar costs £32 and will surface 120 m^2 of roadway. Find the cost of surfacing 1 km of road 18 m wide.
4. A pavement is 600 m long and 4 m wide. It is paved with paving stones each measuring 80 cm by 60 cm and costing £4. Find the cost of the paving stones.
5. A drawing-room window is curtained with material costing £4.25 per metre length. The material comes in strips 75 cm wide. Two curtains are needed, each 1.80 m long and 2.25 m wide. Find the cost of the curtain material.
6. Sailcloth costs £7.20 per m^2. Find the cost of making a sail in the shape of a right-angled triangle with base 2.5 m and height 5 m.
7. Magnetic tape (for recording) costs £1.50 per m^2. The tape in a C60 cassette is 75 m long and 4 mm wide. Find the cost of the tape in a C60 cassette.
8. A reception room is 12 m long and 8 m wide. Round the edge of the room are laid strips of foam-backed carpet 1 m wide, costing £4 per m^2. The rest of the room is carpeted with an Axminster costing £12 per m^2. Find
   (**a**) the size of the Axminster carpet
   (**b**) the cost of carpeting the room in this way.
9. Repeat Question 8 for a room 6 m long and 5 m wide.
10. A new garden is laid out with a lawn measuring 20 m by 15 m, surrounded by a path 2 m wide. Turf for the lawn costs £2.50 per m^2 and gravel for the path costs £0.80 per m^2. Find the cost of turfing the lawn and gravelling the path.
11. A closed cardboard box measures 80 cm by 50 cm by 60 cm. The cardboard costs 50p per m^2. Find the cost of the box.
12. Another closed box measures 20 cm by 10 cm by 15 cm. It is made of cardboard costing 60p per m^2. Find the cost of the box, to the nearest penny.
13. A brick measures 25 cm $\times$ 10 cm $\times$ 8 cm. Each cm^3 has mass 4 g. Find the mass of the brick in kg.
14. Find the mass of a paving stone 80 cm long, 60 cm wide and 5 cm thick, if 1 cm^3 has mass 4 g.
15. Find the mass of a wooden plank 3 m long, 20 cm wide and 3 cm thick, if each cm^3 of wood has mass 0.8 g.
16. Find the capacity in litres of a carton measuring 5 cm by 10 cm by 15 cm.
17. A water tank has length 60 cm, width 45 cm and depth 40 cm. How may litres does it hold when full?
18. A swimming pool is 12 m long, 8 m wide and 8 m deep. How many litres of water are needed to fill it?
19. A water tank measures 80 cm $\times$ 60 cm $\times$ 50 cm. How many litres will it hold when it is three quarters full?
20. Milk costs 54p per litre. It is sold in rectangular plastic containers measuring 15 cm $\times$ 15 cm $\times$ 20 cm. Find the cost of a container full of milk if the empty container costs 5p.
21. 1800 cm^3 of resin is poured into a rectangular tub whose base measures 20 cm by 12 cm. How deep is the resin?
22. An oil storage tank is designed to hold 2250 litres. Its base is 15 m long and 1.2 m wide. How tall must the tank be?

25

Conversion and travel graphs

Exercise 25.1 *Conversion graphs*

1. The total mark for an exam is 60. The teacher wishes to turn the marks into percentages. By taking 60 marks as 100%, draw a graph that could be used to do this conversion. Use a scale 1 cm to 5 marks horizontally and 1 cm to 5% vertically.
   (**a**) Use your graph to find the following marks as percentages:
   (**i**) 30 marks (**ii**) 45 marks (**iii**) 20 marks (**iv**) 54 marks.
   (**b**) Find the marks that gave the following percentages:
   (**i**) 20% (**ii**) 60% (**iii**) 85% (**iv**) $66\frac{2}{3}$%
   (**c**) If the range of the original marks was from 14 to 56, find the range in percentages to the nearest integer.
2. Given that 5 miles is equivalent to 8 km, draw a graph converting miles to kilometres. Using a horizontal scale of 2 cm to 1 km and a vertical scale of 4 cm to 1 mile.
   (**a**) Use your graph to find how many miles are equivalent to each of these:
   (**i**) 4 km (**ii**) 6.4 km (**iii**) 7 km
   (**b**) Use your graph to find how many kilometres are equivalent to each of these:
   (**i**) 3.1 miles (**ii**) 0.6 miles (**iii**) 4.6 miles
   (**c**) Is it possible to use the graph to find the speed in kilometres per hour equivalent to 40 miles per hour?
3. Given that 5 litres of paint will cover 60 m^2, draw a graph that can be used to find how much paint will be required to paint areas up to 60 m^2. Take 2 cm to represent 10 m^2 on the horizontal axis and 4 cm to represent 1 litre on the vertical axis.
   (**a**) Use the graph to find the area that can be painted by each of the following quantities of paint:
   (**i**) 2 litres (**ii**) 3.1 litres (**iii**) 4.5 litres
   (**b**) Use the graph to find *exactly* how much paint will be required to paint the following areas:
   (**i**) 20 m^2 (**ii**) 50 m^2 (**iii**) 3.6 m^2
   (**c**) How many litre tins of paint would be required to paint (**i**) just the walls, (**ii**) the walls and ceiling of a room that is 5 metres long, 4 metres wide and 2.5 metres high?
4. Given that 200 kilometres per hour (km/h) is equivalent to 111 metres per second (m/s) draw a graph enabling you to convert readily from one set of units to the other for speeds up to 200 km/h. Use a horizontal scale of 1 cm to 10 m/s and a vertical scale of 1 cm to 20 km/h.
   (**a**) Use the graph to find the speeds in m/s equivalent to
   (**i**) 28 km/h (**ii**) 60 km/h (**iii**) 96 km/h
   (**iv**) 140 km/h (**v**) 190 km/h

(b) Use the graph to find the speeds in km/h equivalent to
(i) 26 m/s (ii) 44 m/s (iii) 66 m/s
(iv) 88 m/s (v) 94 m/s

5. Given that the current rate of exchange is one pound sterling to $1.90, draw a conversion graph to convert dollars to pounds sterling. Use a scale of 1 cm to 0.5 dollars horizontally and 1 cm to 20 pence vertically.
(a) What would be the cost in dollars of a meal costing £2.60?
(b) How many pounds sterling would be equivalent to $5.70?
(c) Which is the better buy, 3 bars of candy at $2.10 or 5 similar bars at £1.75?

6. The raw marks for the end of term exam range from 93 to 11. (Raw marks are the actual marks obtained out of the total.) These raw marks have to be scaled so that the highest mark is equivalent to 80% and the lowest 30%.
(a) What percentages would be equivalent to raw marks of
(i) 93 marks (ii) 11 marks?
Use this result to draw a graph that can be used to convert the raw marks into equivalent percentages within the scale given. Take a horizontal scale of 2 cm to represent 10% and a vertical scale of 2 cm to represent 10 raw marks.
(b) What percentage be equivalent to each of the following raw marks?
(i) 36 (ii) 44 (iii) 62 (iv) 77
(c) What raw marks would give the following percentages?
(i) 36% (ii) 54% (iii) 65% (iv) 75%
(d) What raw mark would give the same numerical percentage when scaled?

7. Given that 1000 cm^3 are equivalent to 61 cubic inches in volume, draw a conversion graph to convert cubic centimetres into the equivalent volume in cubic inches. Use a vertical scale of 2 cm to 100 cm^3 and a horizontal scale of 2 cm to 10 cubic inches.
(a) Use the graph to convert these into cubic inches:
(i) 590 cm^3 (ii) 280 cm^3 (iii) 835 cm^3
(b) Use the graph to convert these into cubic centimetres:
(i) 31 cubic inches (ii) 20 cubic inches (iii) 46 cubic inches
(c) Will a cuboid box with internal dimensions 7 cm by 11 cm by 13 cm hold 62 cubic inches of sand or not? (The sand must be level with the sides – not heaped.)
(d) Will a rectangular box with internal dimensions 2 inches by 3 inches by 8 inches hold 790 cm^3 of water or not? Give reasons for your answer.

8. An electricity bill is made up of a standing charge of £7.50 and a further charge of 4.5p for every unit of electricity used.
As an example, if 800 units of electricity are used the bill is made up of the £7.50 standing charge + 800 × 4.5p. This gives a total of £43.50.
(a) What would be the total cost if no units of electricity were used?
Use this information to draw a graph to enable you to find the total charge for any number of units up to 800. Take a horizontal scale of 2 cm to 100 units and a vertical scale of 2 cm to £5. Use your graph to find the answers to the following questions:
(b) What would be the total cost for (i) 250 (ii) 700 (iii) 120 (iv) 475 units?
(c) What number of units would give a total cost of
(i) £25 (ii) £32.50 (iii) £10 (iv) £18?
(d) Mr Smithson was charged £36 for 625 units. Was this too much or too little, and by how much?

Exercise 25.2 *Travel graphs*

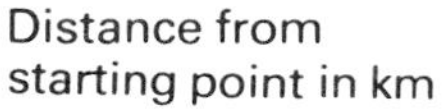

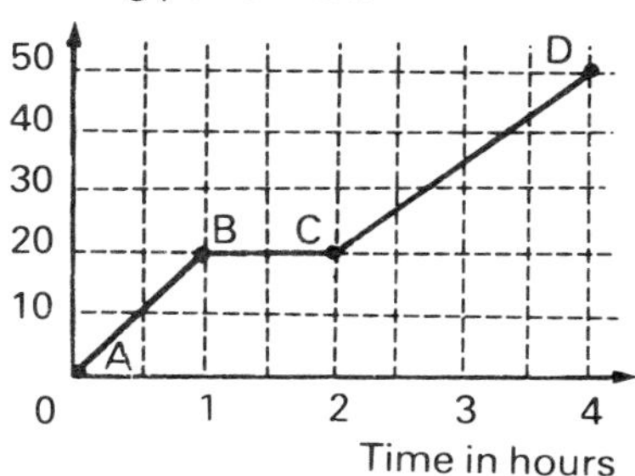

1. The graph on the right illustrates a cycle ride made by a boy to a town 50 km away.
   (**a**) How far and for how long does he travel in section AB of his journey?
   (**b**) What is his speed for that part of the journey?
   (**c**) What do you think happened in the section BC?
   (**d**) How long does his whole journey take?
   (**e**) How far has he travelled altogether?
   (**f**) What is his average speed for the whole journey?

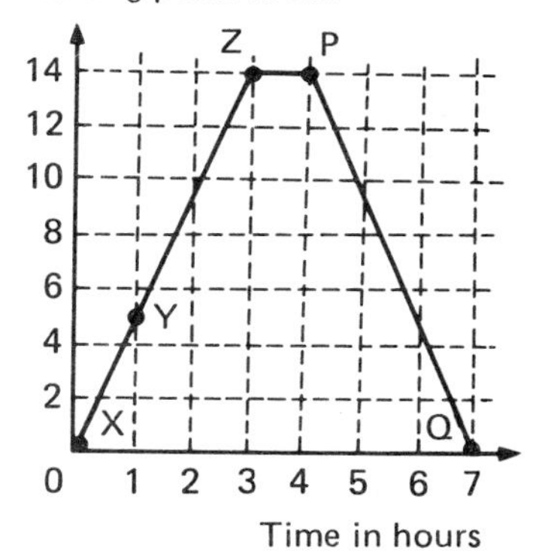

2. The graph on the right illustrates a long walk taken by a girl on holiday.
   (**a**) How far does she travel in section XY?
   (**b**) How far did she travel altogether? (Be careful here!)
   (**c**) For how long did she stop?
   (**d**) What is her speed in km/h on her way home (section PQ)?
   (**e**) How long does her journey take altogether?
   (**f**) What is her average speed for the whole journey?

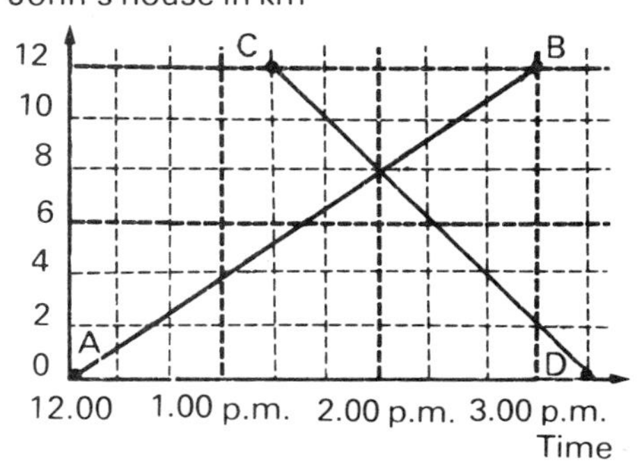

3. Two boys live 12 km apart. At twelve o'clock John starts out to walk to Tom's house. Due to a misunderstanding, Tom starts to walk to John's at 1.20 p.m. Because of thick fog they miss each other, although on the same road, and finally arrive at each other's houses.
   (**a**) Which graph describes John's journey?
   (**b**) Which boy travelled at the faster speed?
   (Don't bother to work the speeds out for this answer.)
   (**c**) At what time did they pass each other?
   (**d**) How far from Tom's house should they have met?
   (Careful here!)

4. Jenny travels to a town 25 km away by bus, has lunch with her grandmother and then travels home by bus.
   (**a**) Which line on the graph describes Jenny's journey?
   (**b**) At what speed was her journey to town?
   (**c**) How long did Jenny spend in town?
   (**d**) The other line shows Charlie walking to his aunt's house with his dog. How far did they go on their walk?
   (**e**) By chance Charlie and his dog were crossing the road just at the times that Jenny was passing in the bus. At what approximate times were these meetings?

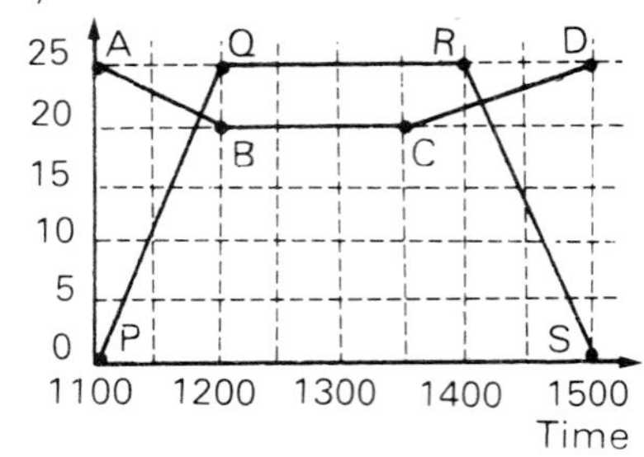

5. Ship P leaves Mars for Earth at exactly the same time that Ship A leaves Earth travelling towards Mars. After $1\frac{1}{2}$ time units (a time unit is 20 days), Ship A reaches an inter- planetary station 1.0×10^7 km from Mars and stays there for a time in order to repair its power unit before returning to Earth. Ship P in the meantime has also arrived at an inter-planetary station and awaits further instructions before eventually continuing to Earth.

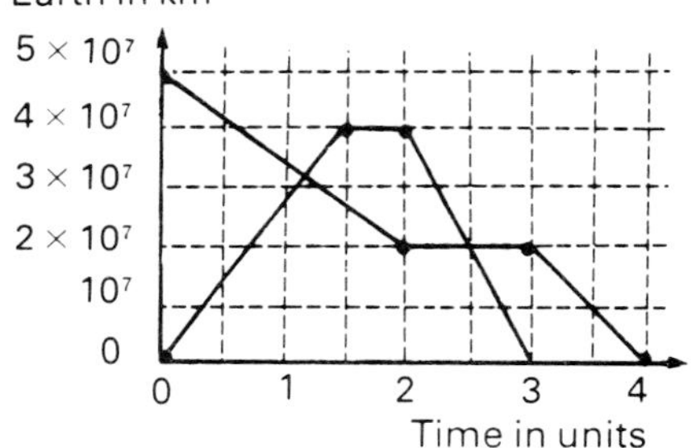

   (**a**) After how many days did Ship P reach its station?
   (**b**) After how many days did Ship A overtake Ship P on the return journey?
   (**c**) How fast was Ship P travelling when overtaken by Ship A on its return journey? Answer in km/day.

6. A plane is thought to have crashed somewhere on its route from Moonville to Beamtown 900 km away. At 10.00 hours the alarm is raised and a helicopter sets out from Moonville to search for signs of wreckage, to be followed at 11.00 hours by a jet from Moonville. The jet quickly overtakes the helicopter and flies on only to be recalled when the helicopter spots the plane which has made a forced landing due to engine trouble. After landing to pick up the passengers and crew of the plane the helicopter returns to base at Moonville.

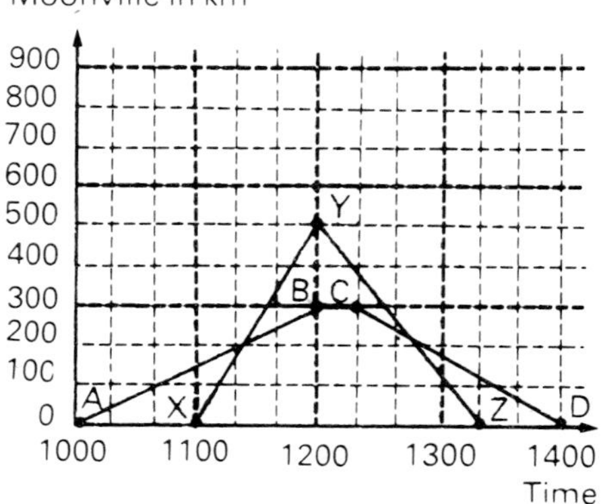

(a) At what time and how far from Moonville did the jet overtake the helicopter?
(b) How far from Beamtown was the plane brought down by engine trouble?
(c) At what time did the helicopter find the plane?
(d) How long did the helicopter stay at the site of the plane?
(e) At what speed was the helicopter travelling when it returned to base?
(f) At what times did the jet fly over the site where the plane landed?

7. A cyclist starts from A at noon and rides steadily at 24 km/h towards B, 60 km away. At 1 p.m. a motorist leaves B for A, travelling at 72 km/h. Draw a graph of these journeys and use it to find out the following:
(a) When did the two meet?
(b) What was their distance from A when they met?

8. A pedestrian sets off to walk along a road at 4 km/h. A cyclist starts from the same place an hour later. She rides along the same road to a town 30 km away at 20 km/h, waits there for half an hour and then returns at the same speed. When and where does she meet the pedestrian on her return journey?

9. Paul goes out on a cross-country ramble. In every hour he walks for twenty minutes, runs for twenty minutes and then rests for twenty minutes. He walks at 4 km/h and runs at 8 km/h. Draw a graph to show Paul's progress for the first 2 hours of his ramble. Show on your graph his progress if he had jogged at 6 km/h for the whole 2 hours. If he had done so, how much further would he have travelled in the 2 hours?

26

Further algebra

Exercise 26.1 *Directed numbers – revision*

1. Point A is 300 metres above sea-level. Point B is 360 metres below A. How far above sea-level is point B?
2. The surface of a certain lake is 640 metres above sea-level and its bottom is −520 metres above sea-level. What is the depth of the lake?
3. The temperature rises from −3°C to 15°C. What is the temperature change?
4. The temperature rises −13°C from 8°C. What will the new temperature be?
5. The temperature changes from 3°C to −5°C. How many degrees has it fallen?
6. Write down four consecutive numbers of which −2 is the least.
7. What is another way of saying 'A clock is −10 minutes fast'?
8. A man buys a bicycle for £24 and sells it for £18. How would you describe his *profit*?
9. What is another way of saying 'My marks are −32 above yours'?
10. What is another way of saying 'I win by −2 metres'?
11. What is another way of saying 'You have lost by −2 goals'?
12. Point B is −32 kilometres due west of Point A. What does this really mean?
13. I turned through −120 degrees clockwise. What does this mean?
14. 'The stone I threw is now −16 metres above me.' What does this really mean?
15. A is 4 kilometres east of B. C is −6 kilometres east of A. How far apart are B and C?
16. A boat is drifting down the river towards London Bridge at a rate of 2 km/h. If it starts 5 kilometres upstream from London Bridge, how far *upstream* from the bridge will it be after (**a**) 1 hour (**b**) 2 hours (**c**) 4 hours?
17. At 1.00 p.m. Peter's watch reads 1.20 p.m. and Ann's watch reads 12.56 p.m. How much *slower* is Ann's watch than Peter's? How much *faster* is Ann's watch than Peter's?
18. A lift starts at the ground floor. It goes up three floors and then goes down one floor. If it then goes up five floors and descends −2 floors, on which floor will it end up?
19. How much greater is 5 than −3?
20. Add together a fall of 23°C and a rise of 15°C and express the result as a *fall* in temperature.

Find the value of each of the following. (If necessary your answer will be in terms of a letter.)

21. $+3a - 2a$
22. $-3a + 2a$
23. $10a - 7a$
24. $-3c - 5c$
25. $-30 - 50$
26. $-17 + 20$
27. $-17n + 20$
28. Subtract +3 from +7.
29. Find the sum of $3d$ and $3d$.
30. Find the difference between $5g$ and $3g$.
31. Show that subtracting 5 is the same as adding −5.

32. Show that the result of adding 3 is the same as subtracting -3.
33. What number added to -5 will make $+4$?
34. What must be added to $+4$ to make the result zero?
35. What must be added to $-4a$ to make the result $+2a$?

Exercise 26.2 *Use of letters – revision*

What expression must be added to:

1. $x+5$ to make $3x+8$?
2. $2x+1$ to make $3x-2$?
3. $4a-3$ to make $2a+1$?
4. $3a+2$ to make $3a-5$?
5. $7-2a$ to make $3a+4$?
6. $3a+4$ to make 9?
7. $2x-1$ to make $a+3$?
8. a^2 to make $a-1$?
9. $3t-5$ to make $t-1$?
10. $7xy+5$ to make $4-xy$?

What expression must be subtracted from:

11. $3a-4$ to give $2a+1$?
12. $2c+1$ to give $3a-2$?
13. $4d+5$ to give $d-2$?
14. $5t-3$ to give $2t+2$?
15. $cd-5$ to give $3cd$?
16. $4t-3s$ to give $2t+s$?
17. $h+3$ to give $-2h+4$?
18. $2t+5$ to give $-t+x$?
19. $7xy-4$ to give $3xy$?
20. $2a+b$ to give $2x-y$?

Collect up like terms in the following:

21. $3b^2-2a^3+b^2-3a^3$
22. $8c^2n+6cn^2-3cn^2+2c^2n$
23. $a^4+a^3+a^2+a+1$
24. $a^2+3a+2a+6$
25. $a^2-3a+2a-6$
26. $8a^2-7ab+5ab-6a^2$
27. $-3x+7x^2-4x^2-2x$
28. $5a^3-7a^2+a^2-6a^3$
29. $x^2y-3xy^2+2x^2y-xy^2$
30. $2x^2+3x^3-5x^5$
31. $y^2-7y^2+12xy^2$
32. $abc+acb-bac-cab$
33. $x^2y+xxy-xyy$
34. $xxx+xx-x$

Simplify the following:

35. $-18d^7 \div 2d^2$
36. $-20x^6 \div 5x^3$
37. $15a^5 \div -3a^3$
38. $\dfrac{-x^2y^2}{xy}$
39. $\dfrac{12e^5f^3}{3af}$
40. $\dfrac{-15m^5n^3}{3m^2n}$
41. $\dfrac{16x^5y^4z^3}{-4xy^3z^2}$
42. $20x^5yz^3 \div 5x^3y^2z^3$
43. $\dfrac{27a^2b^4}{9ab^7}$
44. $\dfrac{3a^4x^2}{-9ax^6}$
45. $\dfrac{16a^5b^3}{(-2ab)^2}$
46. $5a^3 \times (-3a)^2$
47. $9ax(-5x)$
48. $5a^3b^2(-2ab^2)$
49. $-6x^2y \times 4xy^2$
50. $\dfrac{-42a^4b^2}{-7a^2b}$
51. $\dfrac{11a^4b^4c}{5a^2b}$
52. $\dfrac{27a^3b^2}{(-3ab)^2}$
53. $\dfrac{27a^4b^3}{(-3ab)^3}$
54. $\dfrac{(2a)^3b^2}{4a^2b^3}$

Exercise 26.3 *Simple revision problems*

1. If 7 is added to two times a number the result is 11 more than the original number. Form an equation for the number and solve it.
2. The lengths of two opposite sides of a rectangle are $(2x - 3)$ cm and $(x + 5)$ cm respectively. Form an equation and solve it to find
   (**a**) the value of x (**b**) the length of the rectangle
   (**c**) the area of the rectangle if its width is $(x - 4)$ cm.
3. The cost of a pencil is 13 pence less than that of a pen. The pen costs one penny more than 5 times as much as the pencil. How much is
   (**a**) a pen (**b**) a pencil?
4. The cost of a child's fare for a particular train journey is 56 pence less than that for an adult. Three children and two adults travel for £7.32. What is the cost for
   (**a**) an adult (**b**) a child?
5. I walk for 4 hours at x km/h, and then travel by bicycle for another 3 hours at three times my walking speed. How far will I have travelled in the 7 hours in terms of x? If my total journey is 65 km what is my walking pace?
6. The two equal sides of an isosceles triangle are $(2x + 7)$ cm and $(3x - 5)$ cm respectively. What is the value of x?
7. Tom has m records. Charlie has 7 more than Tom and Stephen has 13 fewer than Tom has. If Charlie has three times as many as Stephen, find the number of records that Tom has.
8. Susan raises p pence for charity. Andrew raises twice as much as Susan and Carol raises twice as much as Andrew. If together they raise £3.57, how much did each raise?
9. A number is equal to 4 times (seven less than itself). What is the number?
10. If the price of a book had been 7 pence more, five times its price then would have equalled seven times 11 pence less than the original price. What was the original price?
11. One number is 13 more than another. If five times the smaller is subtracted from three times the larger the result is 11. Find both numbers.
12. A particular polygon with n sides can be divided in a certain way into $(n - 2)$ triangles. If the sum of the angles of all these triangles is equal to the sum of the interior angles of the polygon, find the number of sides when the angle sum is $1440°$.

Exercise 26.4 *Formulation and solution of problems*

In each question, first answer parts (a), (b), etc., then solve the problem.

1. *Problem:* If two numbers add up to 63 and the larger is 15 more than the smaller, what are the two numbers?
   (**a**) If the smaller is x, what is the larger in terms of x?
   (**b**) What do they add up to in terms of x?
   (**c**) Given their total of 63, form an equation and use it to find x.
   (**d**) Having found x, what is the value of the larger number?
   (**e**) Do they add up to 63?
2. *Problem:* If two consecutive numbers add up to 35, what are they?
   (**a**) If the smaller is n, what is the larger in terms of n?
   (**b**) What is their sum, in terms of n?
   (**c**) Since their sum is 35, form an equation for finding n.

(d) What is the larger number?
(e) Check that they add up to 35.

3. *Problem:* If the unequal angle of an isosceles triangle is 100°, what is the size of one of the equal angles?
   (a) If one of the equal angles is x degrees, what will the two angles that are equal add up to in terms of x?
   (b) What will the three angles add up to in terms of x?
   (c) Given that the angles of a triangle add up to 180°, form an equation and use it to find the value of x.
   (d) Do the three angles then add up to 180°?
4. *Problem:* My father is 24 years older than I am. The sum of our ages is 52 years. What will be my father's age in 6 years' time?
   (a) If I am t years old now, what is my father's age now in terms of t?
   (b) What is the sum of our ages, in terms of t?
   (c) Given that the sum of our ages is 52, form an equation and use it to find t, my age now.
   (d) Use t to find my father's age now, and his age in 6 years' time.
5. *Problem:* Jean has 23 more badges than Tom. If together they have 79, how many does each have?
   (a) If Tom has b badges, how many has Jean in terms of b?
   (b) How many do they have altogether, in terms of b?
   (c) Given their total of 79, form an equation for finding b, and hence find how many badges each has.
6. *Problem:* The length of a rectangle is 3 centimetres more than its breadth. If its perimeter is 30 cm, find its length and breadth.
   (a) If the length is d centimetres, what is its breadth in terms of d?
   (b) What would be the perimeter in terms of d?
   (c) Given the perimeter of 30 cm, form an equation and use it to find the length of the rectangle.
   (d) What would be its width or breadth?
   (e) Try to form an equation starting with the width as w centimetres.
7. *Problem:* A dinosaur has an overall length of 27 metres. Its head is $7\frac{1}{2}$ metres shorter than its body and its tail is 3 metres longer than its body. What is the length of its head?
   (a) Let the length of its head be h metres. What is the length of its body in terms of h?
   (b) What is the length of its tail in terms of h?
   (c) What is the total length in terms of h?
   (d) Given the total length of 27 metres, form an equation for h and solve it to find the length of the dinosaur's head.
8. *Problem:* Tom buys 3 bars of chocolate at x pence each and another 4 bars at a shop where they were each 2 pence cheaper. If he had 82 pence change from £2, what is the value of x?
   (a) How much did Tom spend at the first shop in terms of x?
   (b) How much did a bar of chocolate cost at the second shop in terms of x?
   (c) How much did Tom spend at the second shop in terms of x?
   (d) How much did Tom spend altogether in terms of x?
   (e) Given the total amount he spent, form an equation for x. Use it to find the value of x and hence the cost of a bar of chocolate.

9. *Problem:* 240 foreign stamps are divided between Richard, Robert and Keith so that Richard has 20 more than Robert and Robert has 20 more than Keith. How many stamps does each receive?
   (a) Let Richard have m stamps. How many will Robert have in terms of m?
   (b) How many will Keith have in terms of m?
   (c) How many will they have altogether in terms of m?
   (d) Form an equation for the total number of stamps. Solve it to find m, and hence the number of stamps that each has.
10. *Problem:* Three times a certain number is as much above 14 as four times the number is below 161. Find the number.
   (a) How much is three times n above 14, in terms of n?
   (b) How much is four times n below 161, in terms of n?
   (c) Form an equation of these amounts and use it to find n.

Exercise 26.5 *Further problems for formulation and solution*

Using the ideas of Exercise 26.4, solve the following problems:

1. The sum of two consecutive numbers is 85. What are the numbers?
2. My present age is y years. In 6 years' time I will be 29. What is my age now?
3. The sum of two consecutive even numbers is 38. What are the two numbers?
4. If one number is 3 times another and their sum is 48, what are the two numbers?
5. Two numbers differ by 39 and their sum is 73. What are the numbers?
6. Two numbers differ by 14 and their sum is 6. What are the numbers?
7. In 11 years' time I will be 3 times as old as my nephew. If my nephew is now 6 years old, how old am I?
8. Find a number which exceeds its fifth part by 140, taking the original number to be $5x$.
9. Divide £1000 between A, B and C so that A may have £240 more than B, and B may have £20 more than C. How much has each received?
10. A's age is double that of B, while 20 years ago A was 36 years older than B. Find the age of each.
11. Find the angles of an isosceles triangle, if each angle at the base is twice the third angle.
12. It costs 8 times as much to buy a house as to furnish it. If together they cost £8100, how much did it cost to furnish the house?
13. The sum of the interior angles of a polygon of n sides is given by $(180n - 360)$ degrees. If the interior angles of a certain polygon add up to 1080 degrees, how many sides does it have?
14. Find two numbers in the ratio 5 : 3 whose difference is 13.
15. The sides of a triangle are in the ratio 3 : 4 : 5. If the sum of the sides is 54 cm, find the length of the shortest side.
16. When I divided a number of sweets among some boys I found that if I gave each boy 6 sweets I had 8 sweets too few, but if I gave 4 to each boy I had 12 remaining. How many boys were there and how many sweets did I have?
17. One number is 5 greater than another. If I treble the smaller the result is 3 greater than if I double the larger. Find the smaller.
18. Two more than 7 times a number is the same as 26 more than 3 times the number. What is the number?
19. A piece of card is 6 cm wider than it is long. If its perimeter is 40 cm what is its length?

Exercise 26.6 *Problems involving brackets*

When a question has parts (a), (b) etc., work through these first in order to solve the problem.

1. I think of a number, add 9 and then double the result. If the final answer is 32, find the number I thought of first.
   (a) Taking the number I thought of as x, what is it with 9 added, in terms of x?
   (b) What would this become when doubled? (Don't multiply out!)
   (c) Form an equation for the final result and use it to find x, the number that I first thought of.

2. I think of a number, double it, add 3, and double the result. If this gives the answer 30, find the number that I thought of first.

3. I add 6 to a certain number, multiply the result by 5, and the final answer is 65. Find the number.

4. One number is four times as big as another. When 6 is added to each the first becomes three times as big as the second. Find the original numbers.
   (a) Letting the smaller number be n, what will the larger number be in terms of n?
   (b) What will each be in terms of n after 6 has been added to each?
   (c) Form an equation for the new numbers after 6 has been added to each and solve it to find the original numbers.

5. The number that is 3 more than x is twice as big as the number that is 3 less than x. Find x.

6. The present ages of Uncle Albert and Charlie are 63 and 19 respectively. In how many years' time will Uncle Albert be three times as old as Charlie?
   (a) Let the number of years' time be x. How old will they each be in x years' time?
   (b) Form an equation for their ages in x years' time. Solve it to find the value of x, the number of years until Uncle Albert is three times as old as Charlie.

7. Sydney has £17 and Ann has £10. How much must Sydney give Ann in order that Ann may have twice as much as Sydney?

8. The number 23 is divided into two parts, so that 5 times one part plus 2 times the other becomes 73. Find the two parts.
   (a) Letting n be one part, what is the other in terms of n?
   (b) Form an equation. Use it to find n, and hence find the two parts into which 23 is divided.

9. John and Tom have 55 pence between them. If John were to give Tom 8 pence John would then have a quarter of what Tom would have. How much did each have to begin with?

10. The perimeter of a room is 42 metres, and the length is 5 metres more than the width. Find the length.

11. A cyclist travels for 2 hours at a certain speed, and for the next 3 hours at 2 km/h less. If he travels 49 km altogether, find the speed that he travels for the first 2 hours.
   (a) Letting the speed for the first 2 hours be x km/h, what would be his speed for the 3-hour part of the journey?
   (b) How far would he travel in the first 2 hours at x km/h?
   (c) How far would he travel in the last 3 hours, using the answer from (a)?
   (d) Form an equation for the total distance travelled and use it to find the speed for the 2-hour period of the journey.

12. A woman is 24 years older than her daughter, and 2 years ago she was four times as old as her daughter. Find their present ages.
13. Find two consecutive numbers such that 5 times the smaller is 9 greater than 3 times the larger.
14. A man is 9 times as old as his son. In 6 years' time he will be 5 times as old as his son. Find their present ages.
15. John has 15 times as much money as Jean. If John gives Jean 12 pence he then has 3 times as much as she has. How much did each have to start with?

Exercise 26.7 *Equivalent fractions*

Complete the following:

1. $\frac{2x}{5} = \frac{\quad}{15}$
2. $\frac{3x}{4} = \frac{\quad}{12}$
3. $\frac{3}{2y} = \frac{9}{\quad}$
4. $\frac{4}{9} = \frac{4a}{\quad}$
5. $x = \frac{\quad}{4}$
6. $\frac{3}{5} = \frac{3x}{\quad}$
7. $\frac{5}{x} = \frac{\quad}{x^2}$
8. $\frac{a}{y} = \frac{ax}{\quad}$
9. $\frac{7a}{6} = \frac{21a}{\quad}$
10. $\frac{5}{2t} = \frac{5z}{\quad}$
11. $\frac{2a}{3b} = \frac{2ay}{\quad}$
12. $\frac{3a}{5} = \frac{\quad}{15}$
13. $\frac{7t}{5} = \frac{\quad}{10}$
14. $\frac{2y}{3} = \frac{\quad}{12}$
15. $\frac{21d}{6} = \frac{\quad}{2}$
16. $\frac{4x}{12} = \frac{\quad}{3}$
17. $\frac{x-2}{3} = \frac{\quad}{6}$
18. $\frac{h+4}{2} = \frac{\quad}{6}$
19. $\frac{2x+1}{3} = \frac{\quad}{9}$
20. $\frac{1}{4}(2-x) = \frac{\quad}{8}$
21. $\frac{a-2}{3} = \frac{4(a-2)}{\quad}$
22. $\frac{1}{2}(y-1) = \frac{3(y-1)}{\quad}$
23. $\frac{3a+5}{3} = \frac{\quad}{6}$
24. $\frac{x-1}{5} = \frac{2x-2}{\quad}$
25. $\frac{3-2x}{4} = \frac{3(3-2x)}{\quad}$
26. $\frac{t+2}{5} = \frac{2t+t^2}{\quad}$
27. $\frac{2x+14}{14} = \frac{x+7}{\quad}$
28. $\frac{h+2}{6} = \frac{4h+8}{\quad}$
29. $\frac{3x-1}{3} = \frac{\quad}{12}$
30. $\frac{at+a}{5a} = \frac{t+1}{\quad}$

Convert these to equivalent fractions with the lowest common denominator:

31. $\frac{x}{2}, \frac{x}{3}$
32. $\frac{a}{4}, \frac{a}{6}$
33. $\frac{h}{3}, \frac{2h}{4}$
34. $\frac{3}{a}, \frac{2}{b}$
35. $\frac{5x}{6}, \frac{2x}{4}$
36. $\frac{x}{a}, \frac{y}{b}$
37. $\frac{4x}{5a}, \frac{6}{10a}$
38. $\frac{2y}{5}, \frac{3}{3}$
39. $\frac{3b}{5}, \frac{2b}{15}$
40. $\frac{3a}{b}, \frac{2b}{a}$
41. $\frac{a+b}{2}, \frac{a-b}{3}$
42. $\frac{2a-b}{4}, \frac{a}{3}$
43. $\frac{x+a}{4}, \frac{x+b}{3}$
44. $\frac{2x}{3}, \frac{x+1}{2}$
45. $\frac{2v+3}{5}, \frac{v-2}{3}$

Exercise 26.8 *Addition and subtraction of simple algebraic fractions*

Simplify the following:

1. $\frac{x}{5}+\frac{x}{10}$ 2. $\frac{x}{2}+\frac{x}{3}$ 3. $\frac{c}{6}+\frac{c}{4}$ 4. $\frac{2a}{3}-\frac{a}{2}$
5. $\frac{3a}{4}-\frac{a}{3}$ 6. $\frac{1}{a}+\frac{1}{b}$ 7. $\frac{2}{a}-\frac{1}{b}$ 8. $\frac{2a}{5x}+\frac{a}{x}$
9. $\frac{7}{y}-\frac{3}{y}$ 10. $\frac{x}{2}+\frac{y}{3}$ 11. $\frac{a}{6}-\frac{b}{4}$ 12. $\frac{c}{3}+\frac{d}{2}-\frac{e}{6}$
13. $x+\frac{2x}{3}$ 14. $\frac{x}{5}-\frac{7y}{15}$ 15. $\frac{2}{a}+\frac{3}{b}$ 16. $\frac{5}{a}+\frac{3}{a^2}$
17. $\frac{3p}{7}-\frac{5q}{8}$ 18. $\frac{5}{x}-\frac{3}{y}$ 19. $\frac{2}{ab}-\frac{7}{bc}$ 20. $\frac{2k}{h^2}-\frac{1}{h}$

Exercise 26.9 *Addition and subtraction of algebraic fractions – numerical denominators*

1. $\frac{2x-1}{3}+\frac{x+2}{3}$ 2. $\frac{x+3}{4}+\frac{2x-2}{4}$ 3. $\frac{x+1}{2}-\frac{x-2}{2}$
4. $\frac{2x-1}{3}+\frac{x-4}{6}$ 5. $\frac{3x-4}{2}+\frac{x+4}{4}$ 6. $\frac{2x+1}{5}+\frac{3x-1}{10}$
7. $\frac{1}{2}(x-1)+\frac{1}{3}(x+2)$ 8. $\frac{2}{5}(x+3)+\frac{1}{2}(x-4)$ 9. $\frac{3}{4}(x-3)-\frac{1}{3}(x-2)$
10. $x-\frac{1}{2}(x-4)$ 11. $\frac{4a+7}{5}+\frac{3a-4}{15}$ 12. $\frac{2x-5}{3}+\frac{x-1}{6}$
13. $\frac{x-1}{2}+\frac{3x-1}{4}$ 14. $\frac{3a-4b}{6}+\frac{2b-5a}{9}$ 15. $\frac{a-2}{2}-\frac{a-3}{3}$
16. $\frac{a-4}{4}-\frac{a-6}{6}$ 17. $\frac{a-3}{5}-\frac{2a+4}{6}$ 18. $\frac{5(2x-6)}{6}-\frac{1}{3}(4x-7)$
19. $7a-\frac{8a-7b}{3}$ 20. $\frac{4a-b}{5}-\frac{2a-3b}{6}$ 21. $\frac{12y-7}{5}-2y$
22. $\frac{5x+4}{3}-\frac{3(2x+1)}{4}$ 23. $\frac{2}{3}(3x-2)+\frac{x+5}{6}$ 24. $\frac{2x-3}{4}+\frac{x+5}{6}$

Exercise 26.10 *Equations and problems with fractions*

Solve the following equations:

1. $\frac{1}{3}x=6$ 2. $\frac{2}{5}x=4$ 3. $\frac{3}{8}x=1$
4. $\frac{7}{5}x=14$ 5. $\frac{3}{4}x+1=7$ 6. $\frac{2}{3}x-2=x$
7. $\frac{1}{4}x+7=2x$ 8. $\frac{3}{2}x+1=x$ 9. $\frac{1}{4}x+1=2$
10. $2\frac{1}{2}x-3=12$ 11. $\frac{x}{2}+\frac{x}{5}=7$ 12. $\frac{x}{2}=10-\frac{x}{3}$
13. $\frac{x}{2}-\frac{x}{4}=3$ 14. $\frac{1}{6}x+\frac{1}{4}x=15$ 15. $\frac{2}{3}x=11-\frac{x}{4}$
16. $\frac{7}{4}x-\frac{5}{6}x=11$ 17. $x-\frac{x}{8}=49$ 18. $x+\frac{x}{2}+\frac{x}{3}=11$
19. $\frac{x}{2}+\frac{x}{3}=31-\frac{x}{5}$ 20. $\frac{x}{10}+1=x-\frac{2}{3}x$

Solve the following equations:

21. $1.2x = 6.0$
22. $13x = 9.1$
23. $\frac{1}{2}x = 0.7$
24. $\frac{3}{4}x = 1.5$
25. $\frac{1}{5}x = 2.4$
26. $\frac{x}{3} = \frac{7}{9}$
27. $\frac{3}{4}x = 2$
28. $2\frac{2}{3}x = 24$
29. $3\frac{1}{4}x = 3.9$
30. $\frac{7}{8}x = 2.1$
31. $1\frac{3}{8}x = 2.2$
32. $8.5x = 1.7$
33. $1.3x = 0.52$
34. $3.2x = 12.8$
35. $2\frac{5}{8}x = 8\frac{2}{5}$
36. $1.25x = 6.75$
37. $2.5x = 8.75$
38. $5\frac{1}{4}x = 8.4$

39. When a sixth of a certain number is added to a quarter of it the result is 15. What is the number?
40. Find a number such that a third of it is 72 greater than a fifth of it.
41. Find a number such that a sixth of it is 10 less than the number.
42. Find a number such that a quarter of it added to a seventh of it comes to 33.
43. A number, its half and its quarter add up to 126 altogether. What is the number?
44. Find a number such that a quarter of (8 more than the number) comes to 29.
45. Twice a number is 6 more than half the number. What is the number?
46. In a school $\frac{1}{4}$ of the pupils are in the lowest group, $\frac{2}{3}$ in the middle group and 10 in the top group. How many pupils are there in the school altogether?
47. How much money has a boy if $\frac{1}{2}$ of it plus a $\frac{1}{3}$ of it comes to 50 pence?
48. P can do a piece of work in 6 hours and Q can do it in 12 hours. How many hours will it take them to do the work together?
49. A tank is one third full of water. After 12 litres have been poured in it is half full. How many litres will the tank contain when it is full?
50. To two thirds of a certain number I add 10. The result is the same if I take half the number and add 12. What is the number?
51. A car travels a journey of x kilometres at an average speed of 48 km/h. Had it travelled at 72 km/h it would have taken half an hour less. What was the distance travelled by the car?
52. A girl cycles a journey at an average speed of 16 km/h. If she had cycled at 12 km/h the journey would have taken 1 hour longer. What was the distance travelled?
53. Find two consecutive numbers such that a quarter of the smaller added to one fifth of the larger makes 11 altogether.
54. Find two consecutive numbers such that one third of the greater is three greater than one fifth of the smaller.
55. A man walks n km at 5 km/h and then another n km at 4 km/h. If the total time for his journey is $2\frac{1}{4}$ hours, how far does he walk altogether?
56. When 3 is subtracted from two fifths of a certain number the result is one third of the number. What is the number?

Exercise 26.11 *Equations and problems with fractions and brackets*

Solve the following equations:

1. $\frac{1}{2}(x - 2) = 4$
2. $\frac{3}{4}(x - 1) = 6$
3. $\frac{1}{4}(2x + 1) = 2\frac{1}{4}$
4. $\frac{2}{3}(x + 5) = 16$
5. $\frac{2}{3}(x - 4) = 4$
6. $\frac{x - 1}{2} = 6$

7. $\dfrac{2x+10}{3} = 16$

8. $\frac{3}{5}(x-1) = 6$

9. $\frac{3}{4}(x+2) = x$

10. $\frac{1}{4}(x-3) = 2$

11. $\dfrac{3x-1}{5} = 2\frac{1}{5}$

12. $\dfrac{2x+5}{3} = 2\frac{1}{2}$

13. $\dfrac{3x-4}{5} = 2\frac{1}{3}$

14. $\frac{2}{5}(x-3) = 7$

15. $\frac{4}{7}(x+5) = 5\frac{1}{3}$

Solve the following equations:

16. $\dfrac{x-1}{2} + \dfrac{x-3}{3} = 1$

17. $\dfrac{x-6}{5} + \dfrac{x-5}{6} = 1$

18. $\dfrac{x-2}{3} + \dfrac{x-5}{4} = 2\frac{1}{6}$

19. $\frac{1}{2}(x+2) + \frac{1}{3}(x-4) = 8$

20. $5 + \dfrac{x+1}{3} = 7$

21. $3 - \dfrac{x-2}{5} = 0$

22. $1 - \frac{3}{4}(x-2) = 2x$

23. $3 - \frac{1}{5}(x+2) = 0$

24. $x - \frac{2}{5}(2x-3) = 3$

25. $\frac{1}{3}(x+1) - \frac{1}{4}(x-5) = 0$

26. $\dfrac{3(x-2)}{5} - \dfrac{4(x-1)}{7} = 5$

27. $\frac{1}{4}x - \frac{1}{3}(25-x) = 1$

28. $\dfrac{x-4}{3} - \dfrac{x-3}{4} = -\frac{1}{4}$

29. $x - \frac{8}{9}x = \frac{2}{3}x - 30$

30. Find the two parts of 16 such that one third of one part plus half the other part is equal to 6.
31. The result of adding 5 to a number and then dividing by 6 is the same as subtracting 1 from the same number and then dividing by 4. Find the number.
32. I think of a number, subtract 3, divide the result by 4 and add 4. If the final result is 5, what was the number?
33. Tom has 10 pence more than Joan, and one fifth of Tom's money is equal to one third of Joan's. How much money has each?
34. A has £4 less than B, and three sevenths of A's money is equal to two fifths of B's. How much has each?
35. Divide the number 81 into two parts such that three fifths of one part is 3 less than five sixths of the other.
36. A man is 28 years old when his son is born. How many years later will the son's age be half his father's?
37. Two cyclists travel the same journey: one takes $3\frac{3}{4}$ hours and the other $4\frac{1}{2}$ hours. If their speeds differed by 2 km/h find the faster speed.
38. A and B start from two towns 6 km apart at the same time, A walking at a steady rate of $4\frac{1}{2}$ km/h and B at the rate of $3\frac{1}{2}$ km/h. After how long will they meet? How far are they from A's starting point when they meet?
39. One and a half times (8 more than a certain number) is equal to 4 times (7 less than that number). What was the original number?
40. Find two numbers which add up to 90 and are such that one third of the smaller is equal to one seventh of the larger.
41. Five times (3 less than a certain number) is equal to two times (6 more than the same number). What is the number?

42. Peter has twice as much money as John. If Peter gives John ten pence, three quarters of what he now has will equal two thirds of what John now has. How much did each have originally?
43. Paul and Susan have 16 coins between them. They have just enough money to enter a disco, which costs them 50p each. If they have only 10p and 5p coins, how many of each do they have?
44. Find two consecutive numbers such that half one of them plus one third of the other is equal to two fifths of the sum of the two.
45. A cyclist rides to a nearby town at an average speed of 10 km/h and returns at a speed of 15 km/h. If his total journey there and back takes 1 hour 24 minutes, how far away was the town?
46. Find three consecutive numbers such that a quarter of the lowest plus a third of the middle number plus a half of the largest number is equal to 23.
47. Divide 15 into two parts such that one quarter of one part is equal to one sixth of the other.

Exercise 26.12 *Harder inequalities*

The statements 'x is greater than 1' and 'x is less than 6' are combined in a single statement that is written as 'x is greater than 1 but less than 6' in words and $1 < x < 6$ using symbols.
Give the mathematical statements for the following, using symbols:

1. x is less than 4 but greater than 0
2. x is greater than 3 but less than 5
3. x is greater than -1 but less than $+1$
4. x is greater than or equal to -2 but less than 0
5. x is greater than -6 but less than or equal to -2
6. x is less than or equal to 6 and is greater than or equal to 2
7. x is less than -7 and greater than -5
8. x is greater than or equal to -2 and less than or equal to -3
9. x is greater than 6 and less than or equal to 8
10. x is less than or equal to -7 and less than -2

Use the combined statements in Questions 1–10 to answer the following:

11. Using the set of integers, write out the solution set to Question 1.
12. Using the set of natural numbers, write out the solution set to Question 2.
13. Using the set of integers, write out the solution set to Question 3.
14. Is there any difference between your answer to Question 4 and the statement $x = -1$, if you are using the set of (**a**) integers (**b**) rational numbers?
15. What difference does the phrase 'or equal to' make to your solution set for Question 5?
16. Is 6 a member of your solution set for Question 6?
17. Using the set of integers, write out the solution set to Question 7. Is -6 in this set?
18. What is the solution set to Question 8 if you are using the set of (**a**) rational numbers (**b**) integers?
19. Using the set of rational numbers, describe the difference between your solution set to Question 9 and the statement 'x is equal to 7 or 8'.

20. What happens when you try to write out the solution set to Question 10? Give a reason why this happens.

Using the symbols for 'less than', 'greater than' or 'equal to', complete the following statements:

21. $\frac{2}{4}$ $\frac{4}{8}$
22. $\frac{1}{3}$ $\frac{1}{2}$
23. $\frac{1}{3}$ $\frac{1}{4}$
24. $1\frac{1}{3}$ 1.3
25. $\frac{22}{7}$ 3.142
26. An obtuse angle 90°
27. A reflex angle 180°
28. An acute angle 90°
29. $\frac{1}{2}+\frac{3}{4}$ $1\frac{1}{4}$
30. $\frac{1}{3}+\frac{3}{4}$ $1\frac{3}{4}$
31. $\frac{1}{9}$ 0.1
32. $\frac{1}{11}$ 0.1
33. $\frac{5}{11}$ $\frac{1}{2}$
34. 26% $\frac{1}{4}$
35. $33\frac{1}{3}$% $\frac{1}{3}$

Find the solution sets for the following inequalities:

36. $x+3>7$
37. $x-6<5$
38. $4+x>13$
39. $2x+3<7$
40. $3x+4\leqslant 19$
41. $7>2x+1$
42. $7<2x+11$
43. $5x-3\geqslant 7$
44. $2x+3<x+7$
45. $3x-2>x-8$
46. $6x+7\leqslant 3x+16$
47. $2x-3\geqslant 4x-9$
48. $7x-5\geqslant 4x-11$
49. $3x-6<2x-4$
50. $2x-6>3x-4$
51. $12x-1\leqslant 5x+6$
52. $3(x+1)>x+7$
53. $4(x-2)<2x$
54. $5x+1<3(x-3)$
55. $7x<6x+3$

Find the solution sets of the following inequalities:

56. $2-x<7$
57. $x-7>2x$
58. $8+x<-2$
59. $8-x>-2$
60. $2x+4\geqslant 3x-1$
61. $2(x-1)\geqslant 3x$
62. $2x-3(x-2)>10$
63. $2x<3(x-2)$
64. $7-x>3$
65. $2(x-1)-3(x+1)\geqslant 7$
66. $2x-3(x-4)>5x$
67. $7x-2\leqslant 8(x-1)$
68. $2x-4(x-1)\leqslant 6-x$
69. $2(x-3)+5>3(x-1)$
70. $7(x-3)<8x-(x-3)$

Exercise 26.13 *Inequalities involving fractions or decimals*

Solve the following inequalities, leaving your answer as an inequality. Then find the greatest or least integer that will satisfy the inequality.

Example: $\frac{3x}{4}<5$ has the solution $x<6\frac{2}{3}$, and the greatest integer value to satisfy the inequality is $x=6$.

1. $\frac{x}{2}<4$
2. $\frac{x}{3}\geqslant 1$
3. $\frac{2x}{3}\leqslant 3$
4. $5x\leqslant\frac{1}{2}$
5. $7x\leqslant 2$
6. $\frac{3x}{4}>2$
7. $\frac{5x}{6}<\frac{1}{2}$
8. $2\frac{1}{3}x<7$
9. $6\frac{1}{4}x\leqslant 100$
10. $2\frac{1}{2}x\geqslant 4$
11. $3\frac{1}{4}x\geqslant 10$
12. $1.2x\geqslant 6.0$
13. $1.4x\leqslant 0.7$
14. $1.6x<4.0$
15. $2.7x<7.2$

Solve the following inequalities, leaving your answer as an inequality. Then find the greatest or least integer that will satisfy the inequality.

16. $\frac{1}{2}x - 1 < 4$	**17.** $\frac{2}{3}x + 2 > 1$	**18.** $\frac{5}{6}x + 1 \leqslant 2$
19. $2\frac{1}{3}x \geqslant x + 2$	**20.** $\frac{3}{4}x + 2 < 0$	**21.** $\frac{2}{3}x - 1 > \frac{1}{2}x + 1$
22. $1\frac{3}{5}x + 2 \geqslant \frac{2}{3}x$	**23.** $x + \frac{3}{4} < \frac{1}{2}x$	**24.** $1.2x + 0.4 \geqslant 1.6$
25. $0.3x + 1.2 \leqslant 3.3$	**26.** $0.02x - 0.6 > 0.1x + 1.2$	**27.** $\frac{7}{6}x + \frac{2}{3} \geqslant 1\frac{1}{2}$
28. $1\frac{1}{3}x - \frac{1}{4} \leqslant \frac{3}{4}x$	**29.** $\frac{4}{5}x + \frac{2}{3} > \frac{1}{4}$	**30.** $1\frac{3}{4}x + \frac{1}{5} < \frac{2}{3}$

Exercise 26.14 *Simplification of fractions*

Simplify the following where possible. (Hint: look for common factors which can be put outside brackets or cancelled.)

1. $a^3 \times \frac{1}{a}$	**2.** $a^3 \div a$	**3.** $4a \div 2$
4. $\frac{1}{2} \times 4a$	**5.** $\frac{3x+6}{3}$	**6.** $\frac{a^2+a}{a}$
7. $(ab + a) \div a$	**8.** $\frac{a^3+a^5}{a^3}$	**9.** $\frac{2x+4x}{3x}$
10. $\frac{2xy+4xy}{6x}$	**11.** $\frac{6xy+9y}{3y}$	**12.** $\frac{ax-by}{ab}$
13. $\frac{a^2+a}{2} \times \frac{1}{a}$	**14.** $\frac{2x}{y} \times \frac{y^2+y}{4x}$	**15.** $\frac{2}{a+1} \times \frac{a^2+a}{4}$
16. $\frac{3a}{2} \times \frac{4}{3a+6a^2}$	**17.** $\frac{a^2-a^3}{1-a}$	**18.** $\frac{a^2+a^3}{a+1}$

Exercise 26.15 *Further substitution*

If $a = 2$, $b = -1$, $c = \frac{2}{3}$, $d = \frac{1}{2}$, $f = 0$, find the values of the following:

1. ab	**2.** a^2b	**3.** ab^2	**4.** $(ab)^2$
5. $c + d$	**6.** $a(b + d)$	**7.** $c(d + f)$	**8.** $f(d + c)$
9. $fd + fc$	**10.** $d(b + c)$	**11.** $a(c + d)$	**12.** $ab \div cd$
13. $d(a + b)$	**14.** $(a + b)^2$	**15.** $a^2 + b^2$	**16.** $(a - d)^2$
17. $a^2 - d^2$	**18.** $(a + d)(a - d)$	**19.** $c^2 - d^2$	**20.** $a^2d - bc$

If $a = \frac{1}{2}$, $b = \frac{1}{3}$, $c = \frac{3}{4}$, find the values of the following:

21. $a + bc$	**22.** $(b + c) \div a$	**23.** $c \div a^2$
24. $a^2 - b^2$	**25.** a^3	**26.** $2a \div 3b$
27. $2a^2 \div 3b^2$	**28.** $(2a)^2 \div (3b)^2$	**29.** $b(a + c)$
30. $\frac{1}{5}(a + c)$	**31.** $bc \div a^4$	**32.** $\frac{a-c}{b}$
33. abc	**34.** $ab + bc + ac$	**35.** $\frac{1}{2}a(b + c)$

36. If $l = 5$, $b = 4$, find the value of A where $A = lb$.
37. If $h = 6$, $a = 2$, $b = 3$, find the value of A where $A = \frac{1}{2}h(a + b)$.
38. If $a = 3$, $b = 2$, $c = 1.5$ find the value of T where $T = ab + ac + bc$.
39. If $a = 1$, $l = 100$, $n = 20$ find the value of S where $S = \frac{1}{2}n(a + l)$.

40. If $a = 3.5$, $b = 4.5$, $h = 1.25$ find the value of A where $A = \dfrac{h(a+b)}{2}$.

41. If $p = \frac{22}{7}$, $r = 3.5$ find the value of A where $A = pr^2$.

42. If $p = \frac{22}{7}$, $r = 3.5$ find the value of C where $C = 2pr$.

43. If $I = 180(n-2)$ find the value of I when (**a**) $n = 12$ (**b**) $n = 32$ (**c**) $n = 8$.

Given that $T = \frac{1}{2}bh$, find the value of T when:

44. $b = 4$, $h = 2.5$

45. $b = 3$, $h = 6.2$

46. $b = 4.2$, $h = 1.5$

47. $b = \frac{4}{5}$, $h = \frac{5}{8}$

48. $b = \frac{3}{5}$, $h = \frac{2}{3}$

49. $b = 0.8$, $h = 0.625$

50. $b = 2\frac{5}{6}$, $h = 1\frac{5}{17}$

51. $b = 16.3$, $h = 7.2$

52. $b = 5.1$, $h = 6.7$

53. $b = 2.02$, $h = 4.1$

Given that $y = mx + c$, find the value of y when:

54. $m = 2$, $x = 3$, $c = 0$

55. $m = 2$, $x = 3$, $c = 2$

56. $m = \frac{1}{2}$, $x = 3$, $c = 2$

57. $m = -\frac{1}{2}$, $x = 2$, $c = 2$

58. $m = \frac{1}{3}$, $x = 3$, $c = -1$

59. $m = -\frac{1}{4}$, $x = -2$, $c = 3$

60. $m = -\frac{3}{4}$, $x = 3$, $c = \frac{1}{4}$

61. $m = -3$, $x = 1$, $c = 4$

62. $m = 0$, $x = 2.5$, $c = 1$

63. $m = 4$, $x = 1$, $c = -4$

Given that $A = \frac{1}{2}h(a+b)$, find the value of A when:

64. $h = 2$, $a = 3$, $b = 4$

65. $h = 2$, $a = 3.4$, $b = 6.6$

66. $h = 2.4$, $a = 3.6$, $b = 3.4$

67. $h = 0.6$, $a = 2.5$, $b = 6.5$

68. $h = 3.4$, $a = 3.1$, $b = 6.9$

69. $h = 2\frac{2}{3}$, $a = 2\frac{1}{4}$, $b = 3\frac{3}{4}$

70. $h = 1\frac{4}{5}$, $b = 2\frac{1}{2}$, $a = 2\frac{3}{4}$

71. $h = 1.8$, $a = 3.75$, $b = 2.5$

72. $h = 1$, $a = 2.46$, $b = 1.54$

73. $h = 6$, $a = 0$, $b = 4.5$

74. Given that $L = a + (n-1)d$, find the value of L when:
(**a**) $a = 1$, $n = 20$, $d = 2$ (**b**) $a = 101$, $n = 11$, $d = 1$
(**c**) $a = 1$, $n = 51$, $d = 2$

75. Given that $S = 2(lb + bh + lh)$, find the value of S when:
(**a**) $l = 2$, $b = 3$, $h = 4$ (**b**) $l = 2.5$, $b = 3$, $h = 4$
(**c**) $l = 2.5$, $b = 3.5$, $h = 1.5$

76. Given that $C = \frac{5}{9}(F - 32)$ find the value of C when:
(**a**) $F = 32$ (**b**) $F = 113$ (**c**) $F = 212$

77. Given that $S = 2pr(r + h)$ where $p = \frac{22}{7}$, find the value of S when:
(**a**) $r = 3.5$, $h = 2$ (**b**) $r = 1.4$, $h = 10$ (**c**) $r = 8.4$, $h = 8$

78. Given that $S = \dfrac{n}{2}(2a + (n-1)d)$ find the value of S when:
(**a**) $a = 1$, $n = 50$, $d = 1$ (**b**) $a = 1$, $n = 10$, $d = 4$
(**c**) $a = 1$, $n = 15$, $d = 3$

79. Given that $D = ut + \frac{1}{2}ft^2$, find the value of D when:
(**a**) $u = 0$, $f = 10$, $t = 20$ (**b**) $u = 20$, $f = -10$, $t = 4$
(**c**) $u = 40$, $f = -10$, $t = 10$

80. Given that $R = p(a - b)(a + b)$ find the value of R when:
(a) $p = \frac{22}{7}$, $a = 10$, $b = 3$ (b) $p = 3.14$, $a = 7$, $b = 3$
(c) $p = 3.14$, $a = 6.4$, $b = 3.6$

Exercise 26.16 *Rearranging formulae*

Change the subject of each of the following formulae so that the letter in brackets is the new subject:

1. $I = bh$ (h)
2. $A = \frac{1}{2}bh$ (b)
3. $S = 180n - 360$ (n)
4. $d = st$ (s)
5. $D = \dfrac{M}{V}$ (V)
6. $V = lbh$ (b)
7. $C = 2pr$ (r)
8. $S = 2prl$ (l)
9. $A = pr^2$ (r)
10. $C = 4pr^2$ (r)
11. $S = \dfrac{PRT}{100}$ (P)
12. $a = \frac{1}{2}h(x + y)$ (h)
13. $a = \frac{1}{2}h(x + y)$ (x)
14. $y = mx + c$ (x)
15. $l = a + (n - 1)d$ (d)
16. $l = a + (n - 1)d$ (n)
17. $F = \frac{9}{5}C + 32$ (C)
18. $S = \frac{1}{2}n(2a + (n - 1)d)$ (a)

Make the given letter the subject of the formula and then find its value using the other values given:

19. $A = bh$ (b) when $h = 3$, $A = 24$
20. $D = \dfrac{M}{V}$ (M) when $D = 1.2$, $V = 15$
21. $D = \dfrac{M}{V}$ (V) when $D = 0.8$, $M = 84$
22. $S = \dfrac{PRT}{100}$ (R) when $P = 10$, $T = 3.5$, $S = 7$
23. $S = \dfrac{PRT}{100}$ (T) when $P = 40$, $R = 18$, $S = 72$
24. $A = \frac{1}{2}h(a + b)$ (h) when $A = 40$, $b = 3.5$, $a = 4.5$
25. $A = \frac{1}{2}h(a + b)$ (a) when $A = 12.3$, $h = 6$, $b = 2.4$
26. $a = \dfrac{c^2}{b}$ (c) when $a = 24$, $b = 6$
27. $P = 2(a + b)$ (a) when $P = 13$, $b = 3.2$
28. $\dfrac{1}{f} = \dfrac{1}{u} + \dfrac{1}{v}$ (v) when $f = 2.4$, $u = 4$

Make the given letter the subject of the formula and then find its value using the other values given:

29. $A = PR^2$ (R) when $A = 314$, $P = 3.14$
30. $A = PR^2$ (R) when $A = 1.54$, $P = \frac{22}{7}$
31. $v = u + ft$ (t) when $v = 120$, $u = 60$, $f = 10$
32. $v = u + ft$ (t) when $v = 40$, $u = 20.4$, $f = 9.8$
33. $P = aW + b$ (W) when $P = 34$, $b = 24.4$, $a = 1.6$

34. $S = 2pr(r + h)$ (h) when $p = \frac{22}{7}$, $r = 3.5$, $S = 176$

35. $v^2 = u^2 + 2as$ (a) when $v = 6$, $u = 4$, $s = 4$

36. $v^2 = u^2 + 2as$ (a) when $v = 6.3$, $u = 3.7$, $s = 5$

37. $D = \dfrac{3h}{2}$ (h) when $D = 4$

38. $P = A + V$ (A) when $P = 3.1$, $V = 3.61$

27

Further work with angles, polygons and solids

Exercise 27.1 *Further angle calculations*

(Give reasons for all statements you make.)

1. ABC is a straight line.
AB = AD
$\widehat{DAB} = 38°$
Calculate (**a**) $\widehat{ADB}$ (**b**) $\widehat{DBC}$.

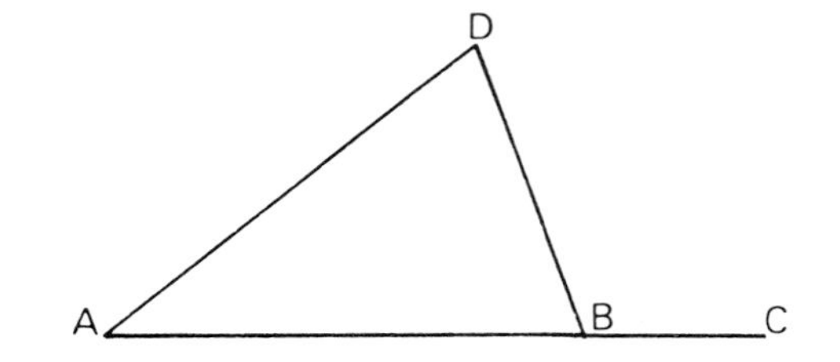

2. ABC and DBE are straight lines.
CB = CE
$\widehat{ABD} = 66°$
Calculate (**a**) $\widehat{EBC}$ (**b**) $\widehat{ECB}$.

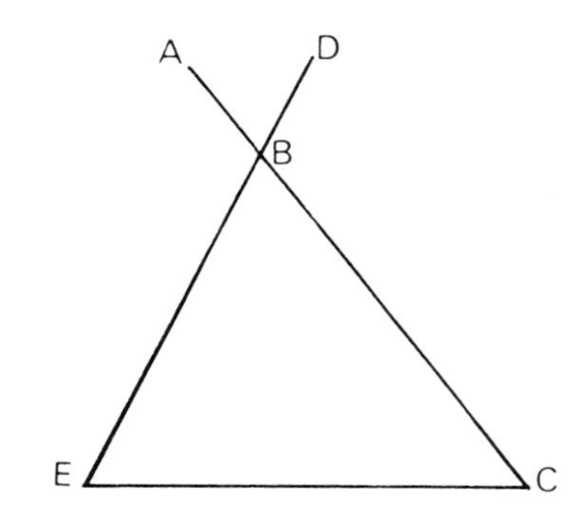

3. PQR and SQT are straight lines.
QP = QT
$\widehat{TQR} = 130°$
Calculate (**a**) $\widehat{RQS}$ (**b**) $\widehat{TQP}$
(**c**) $\widehat{TPQ}$.

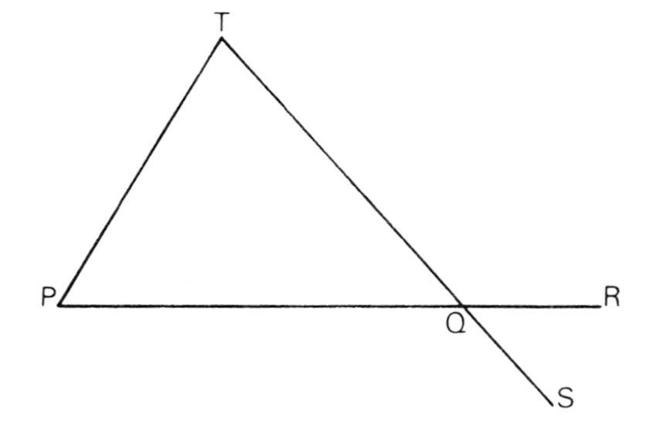

4. CD bisects the reflex angle ACB.
CA = CB
$\widehat{ACD} = 125°$
Calculate (**a**) $\widehat{ACB}$ (**b**) $\widehat{ABC}$.

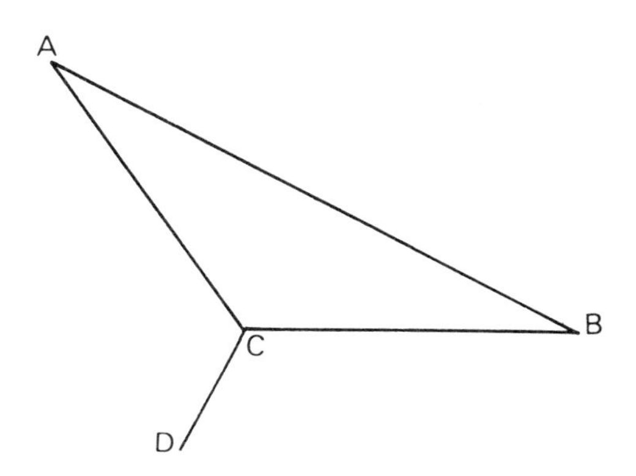

5. ACD and BCE are straight lines.
ABC is a right-angled isosceles triangle with the right angle at A.
$\widehat{CDE} = 40°$
Calculate (**a**) $\widehat{ACB}$ (**b**) $\widehat{DCE}$
(**c**) $\widehat{CED}$.

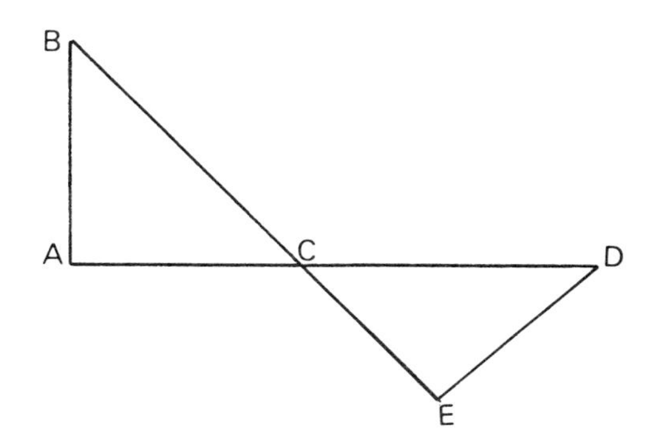

6. BA = BC and AC = AD
$\widehat{ABD} = 30°$
Calculate (**a**) $\widehat{ACB}$ (**b**) $\widehat{CAD}$
(**c**) $\widehat{DAB}$ (**d**) $\widehat{BDA}$.

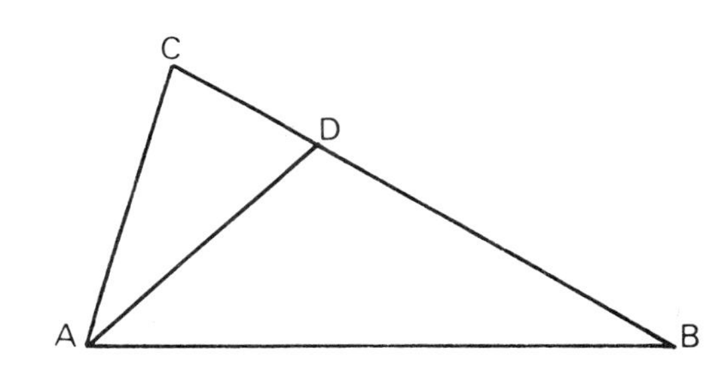

7. CA = CB = AD
$\widehat{ABC} = 52°$
$\widehat{BAD} = 100°$
Calculate (**a**) $\widehat{BAC}$ (**b**) $\widehat{CAD}$
(**c**) $\widehat{ACD}$ (**d**) $\widehat{BCD}$.

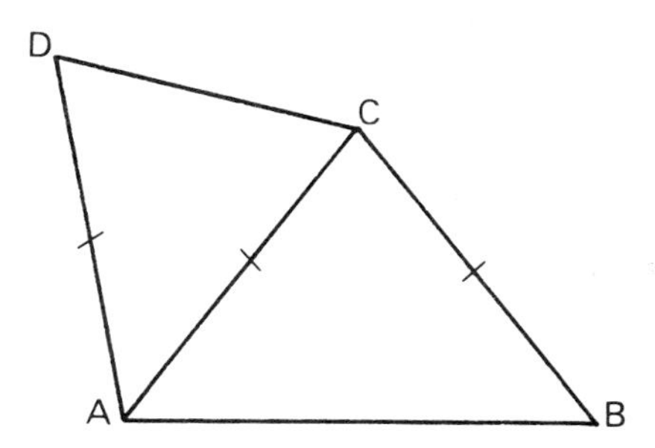

8. ABCD is a rectangle.
$\widehat{DOC} = 110°$
Calculate (**a**) $\widehat{COB}$ (**b**) $\widehat{OCB}$
(**c**) $\widehat{OCD}$.

9. OA, OB and OC are radii of the circle.
$\widehat{AOB} = 60°$
(**a**) What kind of triangle is AOB?
(**b**) Calculate $\widehat{BOC}$ and $\widehat{OBC}$.

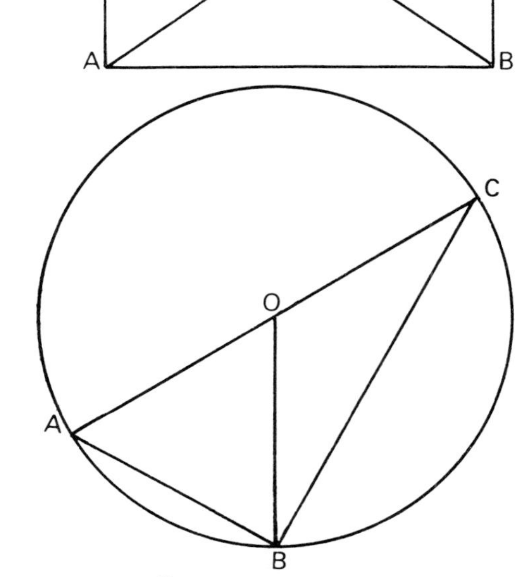

10. ABCDE is a regular pentagon.
O is the centre of the figure.
(**a**) What is the size of $\widehat{AOB}$? (Think of the rotational symmetry of the figure.)
(**b**) What do you know about OA and OB?
(**c**) Calculate $\widehat{OAB}$.
(**d**) What does OA do to $\widehat{BAE}$?
(**e**) Calculate $\widehat{BAE}$.
(**f**) Calculate $\widehat{EDA}$.

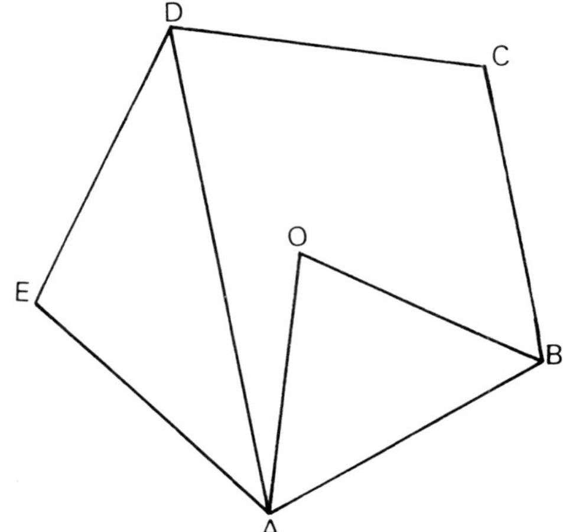

Exercise 27.2 *Miscellaneous harder angle calculations*

(Give reasons for statements you make.)

1. AB is parallel to CD.
CB = CD
$\widehat{ABC} = 36°$
Calculate (**a**) $\widehat{BCD}$ (**b**) $\widehat{CDB}$.

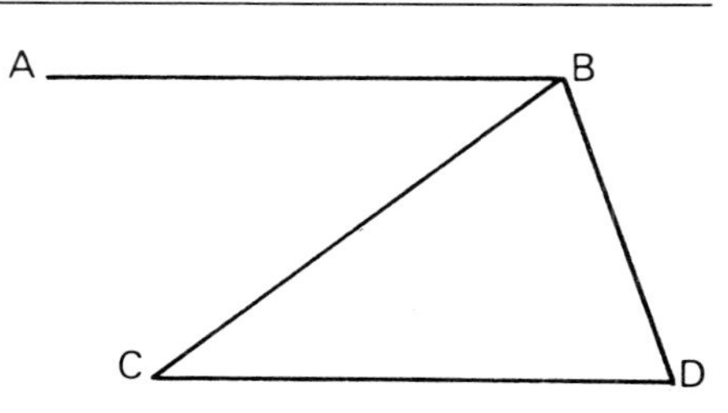

2. PQ is parallel to RS.
$QR = QS$
$\widehat{PQR} = 38^\circ$
Calculate (**a**) $\widehat{QRS}$ (**b**) $\widehat{RSQ}$.

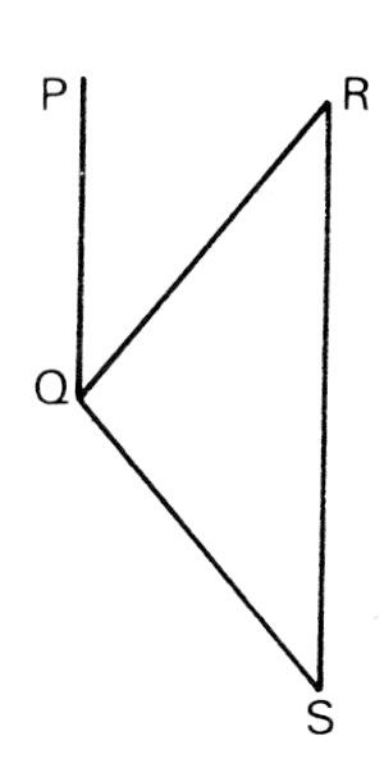

3. AB is parallel to EC.
$CB = CD$
$\widehat{ABD} = 48^\circ$
Calculate (**a**) $\widehat{BDC}$ (**b**) $\widehat{BCD}$.

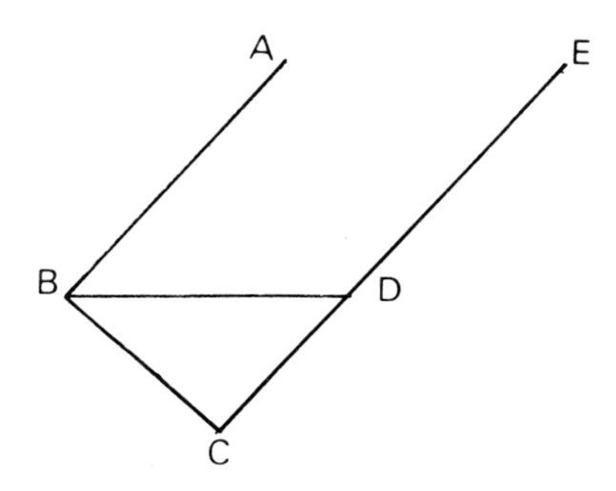

4. OP is parallel to QR.
$RS = RT$
$\widehat{QTS} = 112^\circ$
Calculate (**a**) $\widehat{STR}$ (**b**) $\widehat{TRS}$
(**c**) $\widehat{POS}$. (**d**) At what angle would TS produced meet OP?

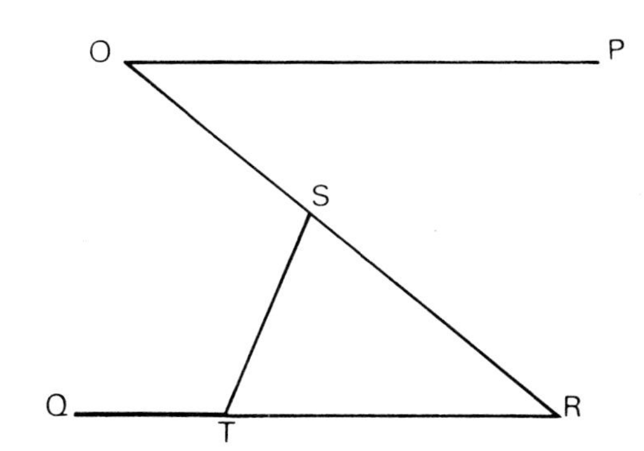

5. ABCD is a rhombus.
$\widehat{BDA} = 75^\circ$
Calculate (**a**) $\widehat{DAB}$ (**b**) $\widehat{ABC}$.

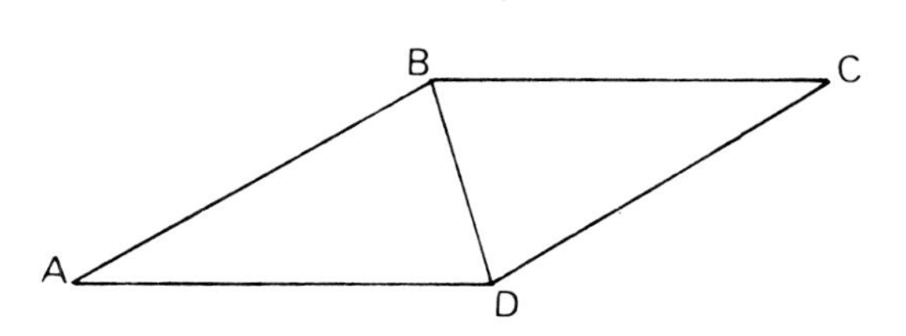

6. WXYZ is a parallelogram.
$X\widehat{W}Y = 30°$
$W\widehat{Z}Y = 105°$
Calculate (a) $Z\widehat{Y}W$ (b) $W\widehat{X}Y$.

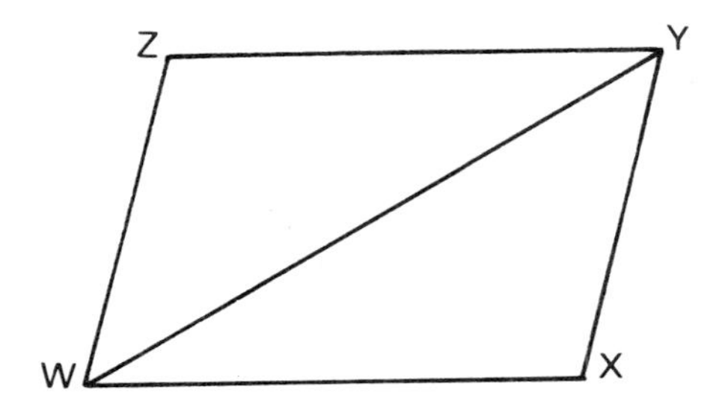

7. AB is parallel to DC.
$OA = OB$
$A\widehat{B}O = 47°$
Calculate (a) $O\widehat{D}C$ (b) $C\widehat{O}D$.
(c) What can you say about $\triangle COD$?

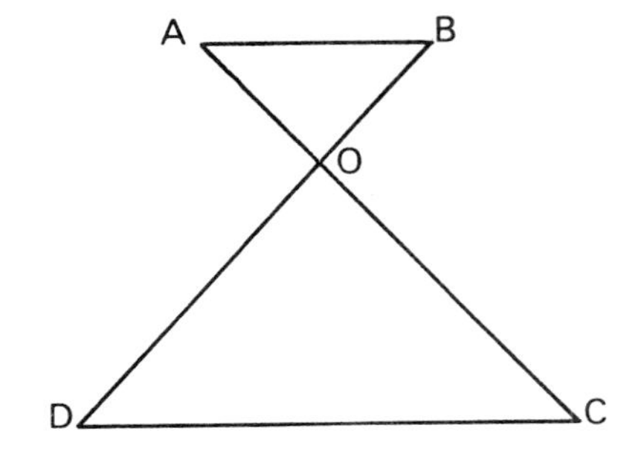

8. PQ is parallel to RS.
$AB = AC$
AC bisects $B\widehat{A}Q$.
$R\widehat{B}A = 68°$
Calculate (a) $B\widehat{A}C$ (b) $A\widehat{C}B$
(c) $C\widehat{B}S$ (d) $P\widehat{A}C$.
(e) What is the angle between AC produced and RS?

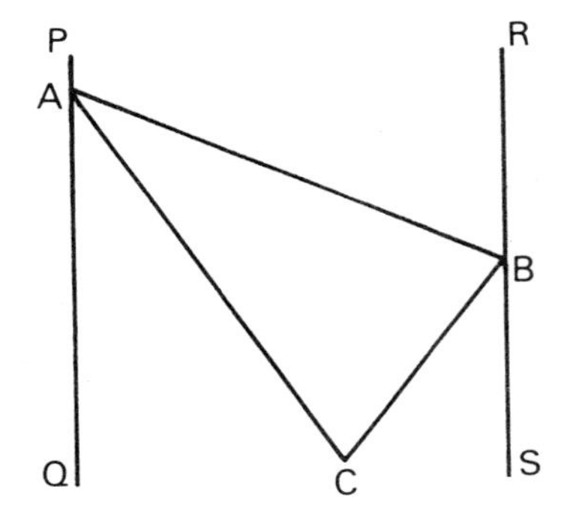

9. AC is parallel to DG.
$BE = BF$
$B\widehat{F}G = 105°$
Calculate (a) $B\widehat{E}F$ (b) $A\widehat{B}E$
(c) $E\widehat{B}C$.

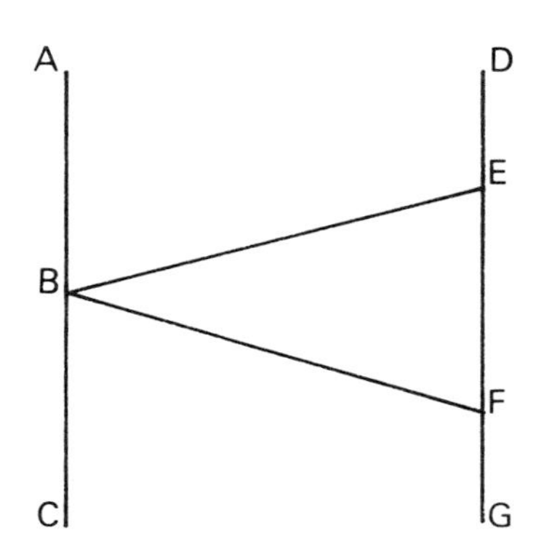

10. PQRS is a trapezium with PQ parallel to RS.
$\widehat{QPS} = 73°$
$\widehat{SRQ} = 127°$
Calculate (**a**) $\widehat{PQR}$ (**b**) $\widehat{PSR}$.
(**c**) At what angle would PS produced and QR produced meet?

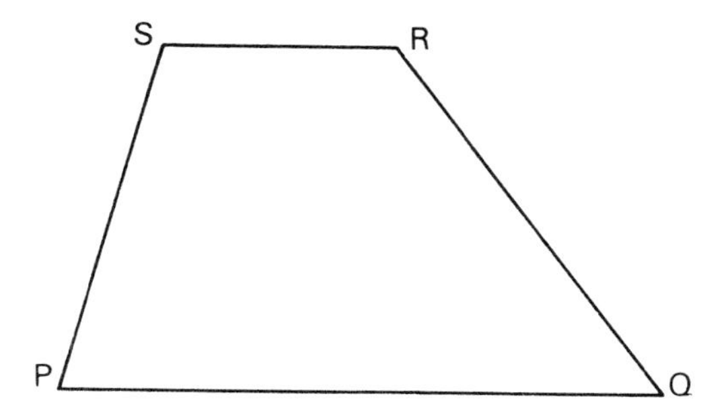

Exercise 27.3 *Construction of quadrilaterals*

The following facts about quadrilaterals will help you construct the figures in this exercise:

Square

1. All sides are equal.
2. All angles are 90°.
3. Diagonals are equal, mediate each other and bisect the angles at the corners.

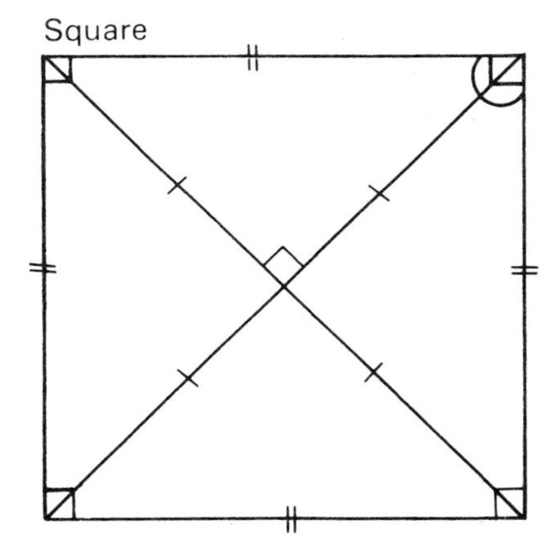

Rectangle

1. Opposite sides are equal.
2. All angles are 90°.
3. Diagonals are equal and bisect each other.

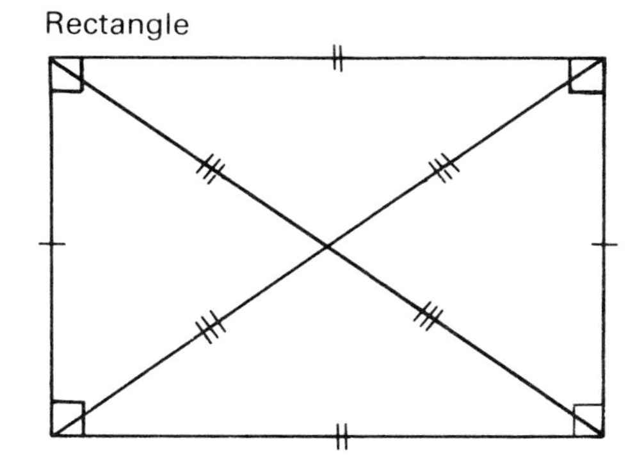

Rhombus

1. All sides are equal.
2. Opposite sides are parallel.
3. Opposite angles are equal.
4. Diagonals mediate each other and bisect the angles at the corners.

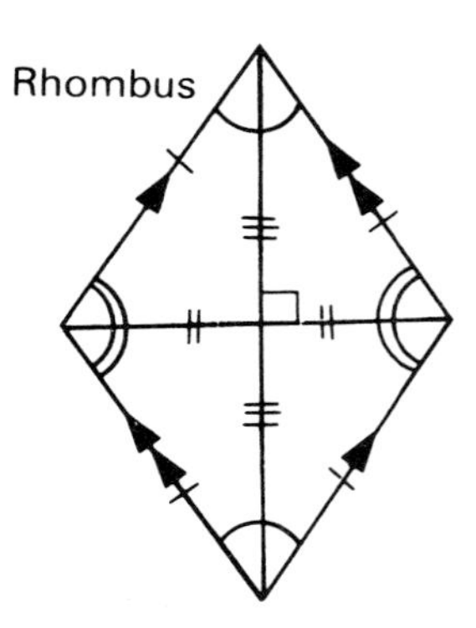

Parallelogram

1. Opposite sides are equal and parallel.
2. Opposite angles are equal.
3. Diagonals bisect each other.

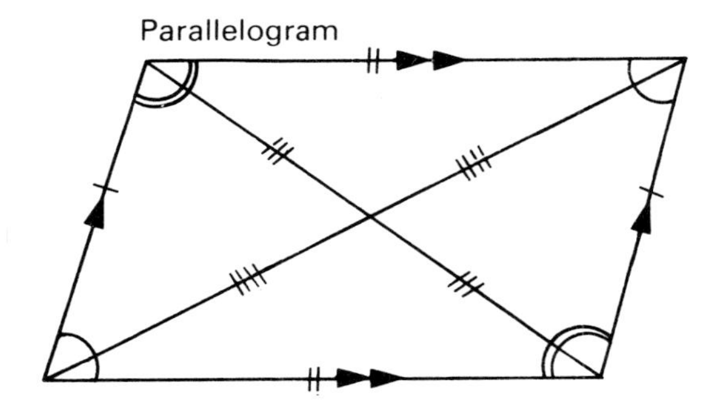

Trapezium This figure has one pair of parallel sides only (but see below).

Isosceles Trapezium

The non-parallel sides and the diagonals are equal. The angles on the parallels are equal.

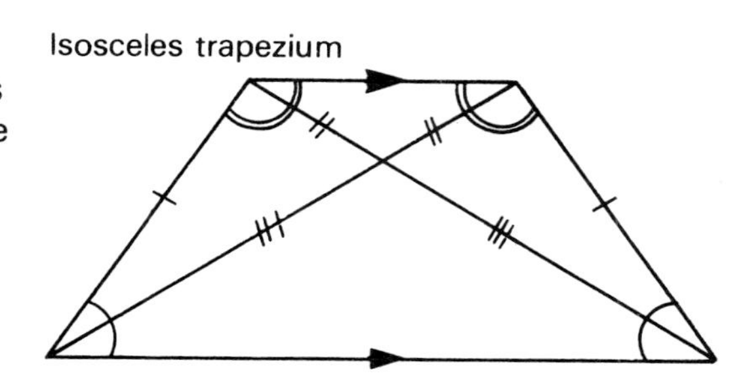

Kite

1. Two pairs of adjacent sides equal.
2. One pair of opposite angles equal.
3. One diagonal mediates the other.
4. One diagonal bisects the opposite angles.

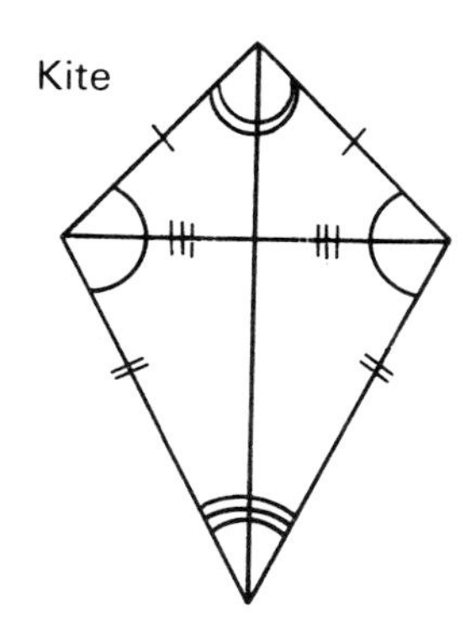

Arrowhead

1. Two pairs of adjacent sides equal.
2. One pair of angles equal.
3. One reflex angle.
4. Diagonal bisects angles.

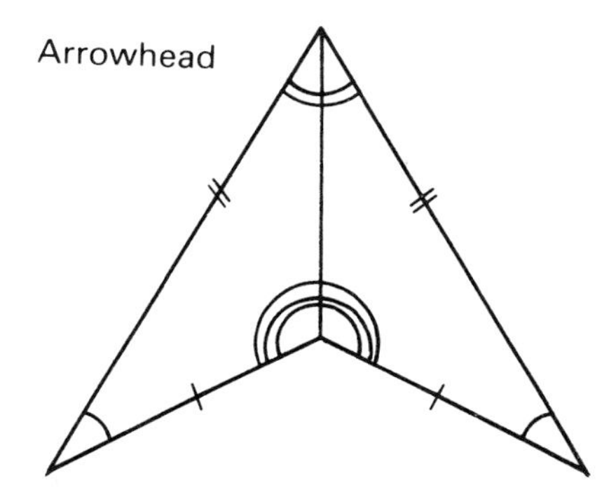

1. Using ruler and compasses only, construct the square ABCD when AB = 6 cm. Measure AC.
2. Using ruler and compasses only, construct the square whose diagonals are 7 cm long. What is the length of a side?
3. In rectangle ABCD, AB = 7 cm and BC = 4 cm. Construct the figure, using ruler and compasses only. Measure AC.
4. Construct rectangle PQRS when PQ = 8 cm and PR = 9 cm. What is the length of SP?
5. The diagonals of a rectangle ABCD are 8.6 cm and the obtuse angle between them is 130°. Construct the figure and measure AB and AD.
6. In the rhombus WXYZ, WX = 5 cm and $\widehat{WXY} = 128°$. Construct the figure and measure XZ.
7. Construct rhombus PQRS when PQ = 4.3 cm and PR = 5 cm. Measure $\widehat{PQR}$.
8. Construct a rhombus whose diagonals are 7 cm and 5 cm, using ruler and compasses only. What is the length of the perimeter?
9. Construct the rhombus ABCD when AC = 6 cm and $\widehat{BAD} = 80°$. What is the length of BD?
10. ABCD is a parallelogram in which AB = 8 cm, BC = 4.5 cm and $\widehat{ABC} = 116°$. Construct the figure and measure the diagonals.
11. Construct parallelogram PQRS when PQ = 7.5 cm, QR = 4 cm and PR = 10.2 cm. Measure $\widehat{PSR}$.
12. Construct parallelogram WXYZ if WX = 7.5 cm, XY = 5.5 cm and $\widehat{WXY} = 52°$. What is the length of WY?

13. The diagonals of parallelogram ABCD meet at O. Construct the figure if $AC = 11$ cm, $BD = 6$ cm and $A\widehat{O}B = 140°$. Measure AB and AD.
14. ABCD is an isosceles trapezium in which $AD = BC$. Construct the figure when $AB = 10$ cm, $AD = 6$ cm and $A\widehat{B}C = 65°$. Measure CD.
15. Construct a kite ABCD when $AB = 6.6$ cm, $BC = 9.6$ cm and $AC = 12$ cm. Measure $B\widehat{A}D$.
16. The diagonals of a kite meet at O. Using ruler and compasses only, construct the kite if $AC = 10$ cm, $BD = 5$ cm and $AO = 4$ cm. What is the length of CD?
17. In the kite PQRS, $PQ = PS = 5.5$ cm, $PR = 12.5$ cm and $Q\widehat{P}S = 65°$. Construct the figure and measure QR.
18. ABCD is an arrowhead. Construct the figure if $AB = 9$ cm, $BC = 6.5$ cm and $A\widehat{B}C = 15°$. Measure AC.
19. Construct the arrowhead PQRS if $PQ = 9.5$ cm, $QR = 4.8$ cm and the axis of symmetry $PR = 5.6$ cm. Measure the reflex angle $Q\widehat{R}S$.
20. Construct the quadrilateral ABCD given that $AB = 7.5$ cm, $BC = 4.5$ cm, $AD = 12$ cm, $A\widehat{B}C = 113°$ and $B\widehat{C}D = 135°$. Measure the lengths of the diagonals.

Exercise 27.4 *Further work with polygons*

1. Make rough copies of the figures given below and then follow the instructions carefully. From one vertex (corner) draw as many diagonals as you can until the figure is divided entirely into triangles. (See the example given for the pentagon). You should then be able to complete the table given below:

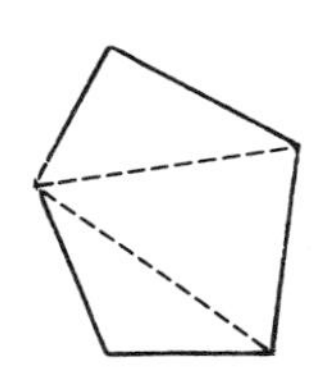 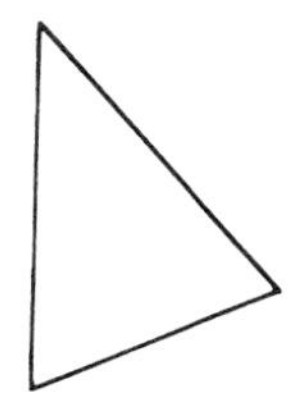  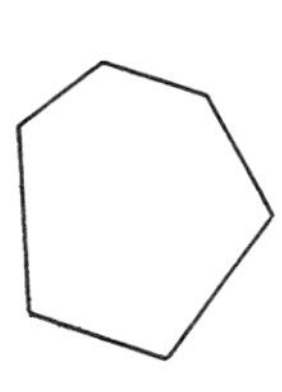 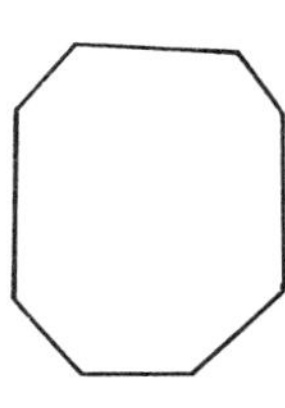 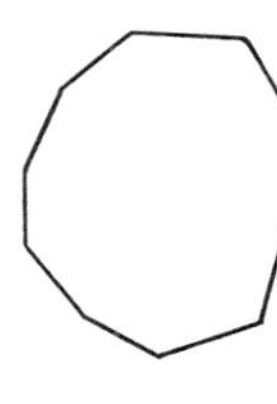

No. of sides	Name of polygon	No. of triangles	Interior angle sum in degrees
5	pentagon	3	$3 \times 180° = 540°$
3	?	?	? = ?
4	?	?	? = ?
6	?	?	? = ?
8	?	?	? = ?
10	?	?	? = ?
n	–	?	? = ?

2. Using your result for the n-sided polygon calculate the interior angle sum for the following number of sides in the polygon:
(**a**) 12 (**b**) 15 (**c**) 22 (**d**) 30 (**e**) 1002 (**f**) 2

3. Work out the size of each interior angle for each of the polygons listed in Question 1, assuming that each is regular.
4. Using the results to Question 1, work out:
   (**a**) the size of each exterior angle,
   (**b**) the sum of the exterior angles,
   in each case.

All the following questions refer to the parallelogram ABCD. What name do we give the figure in each case if the following facts about it are true? (Consider each question separately from the others.)

5. Angle BAC = angle ACB
6. Angle BAC + angle ACB = 90°
7. Angle BAC = angle ACB = 45°
8. Angle BOC = 90°
9. AC = BD
10. Angle BOC = 90° and AC = BD
11. OA = OB and angle OBA = 45°
12. Triangle OAB is equilateral.
13. A, B, C, D all lie on a circle centre O.
14. What name do we give the figure if we know that it is not a parallelogram; AD is parallel to BC but AB is not parallel to DC?

Exercise 27.5 *Problems with polygons*

1. Find the third angle of a triangle whose other two angles are
   (**a**) 63°, 70° (**b**) 40°, 100° (**c**) 90°, 45° (**d**) 60°, 60°
   Give the special names for the triangles of (b), (c) and (d).
2. Find the fourth angle of a quadrilateral whose other three angles are
   (**a**) 84°, 143°, 75° (**b**) 36°, 40°, 144° (**c**) 70°, 70°, 110°
   Give the special names for the quadrilaterals of (b) and (c). (There are two answers to (c).)
3. (**a**) What is the sum of the exterior angles of any polygon that is regular? Use this fact to find the size of an exterior angle in a regular polygon with these numbers of sides:
   (**b**) 18 (**c**) 15 (**d**) 20
4. Using your answers to Question 3, find the size of an interior angle in a regular polygon with these numbers of sides:
   (**a**) 18 (**b**) 15 (**c**) 20 (**d**) 30 (**e**) 90
5. How many sides has a polygon in which the sum of the interior angles is exactly twice the sum of the exterior angles?
6. Is it possible for a regular polygon to have interior angles of 130°? Give a reason for your answer.
7. How many sides has a polygon in which the sum of the interior angles is exactly three times the sum of the exterior angles?
8. How many sides has a polygon in which the sum of the interior angles is exactly four times the sum of the exterior angles?

Exercise 27.6 *Solid figures*

Solid shapes of congruent cross-section

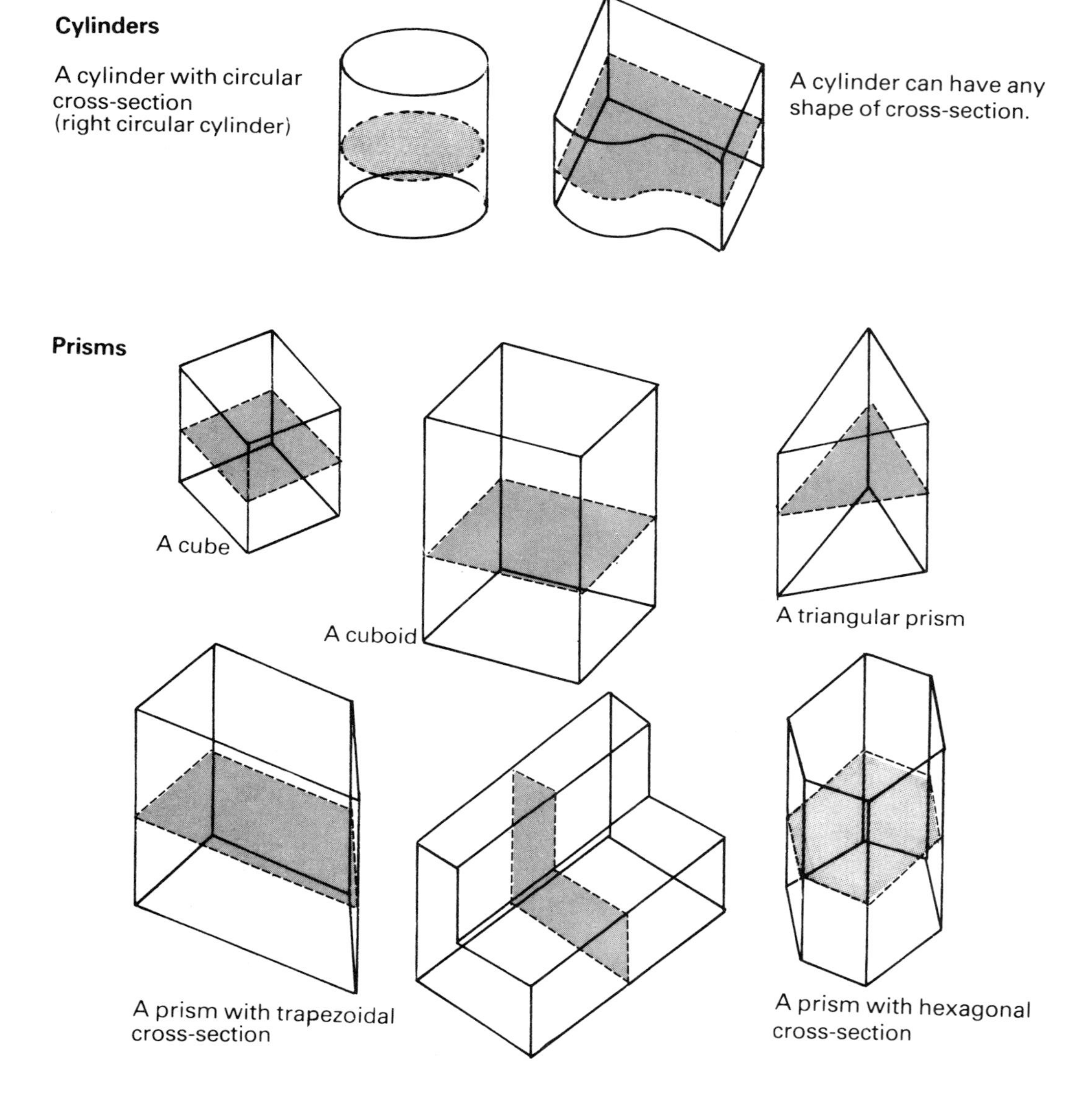

A prism can have any polygon as its cross-section. This cross-section is taken parallel to the base and the top which are themselves parallel to each other. The cross-section must be exactly the same shape and size (congruent) throughout the prism. (Note the shaded cross-sections.)

Cones

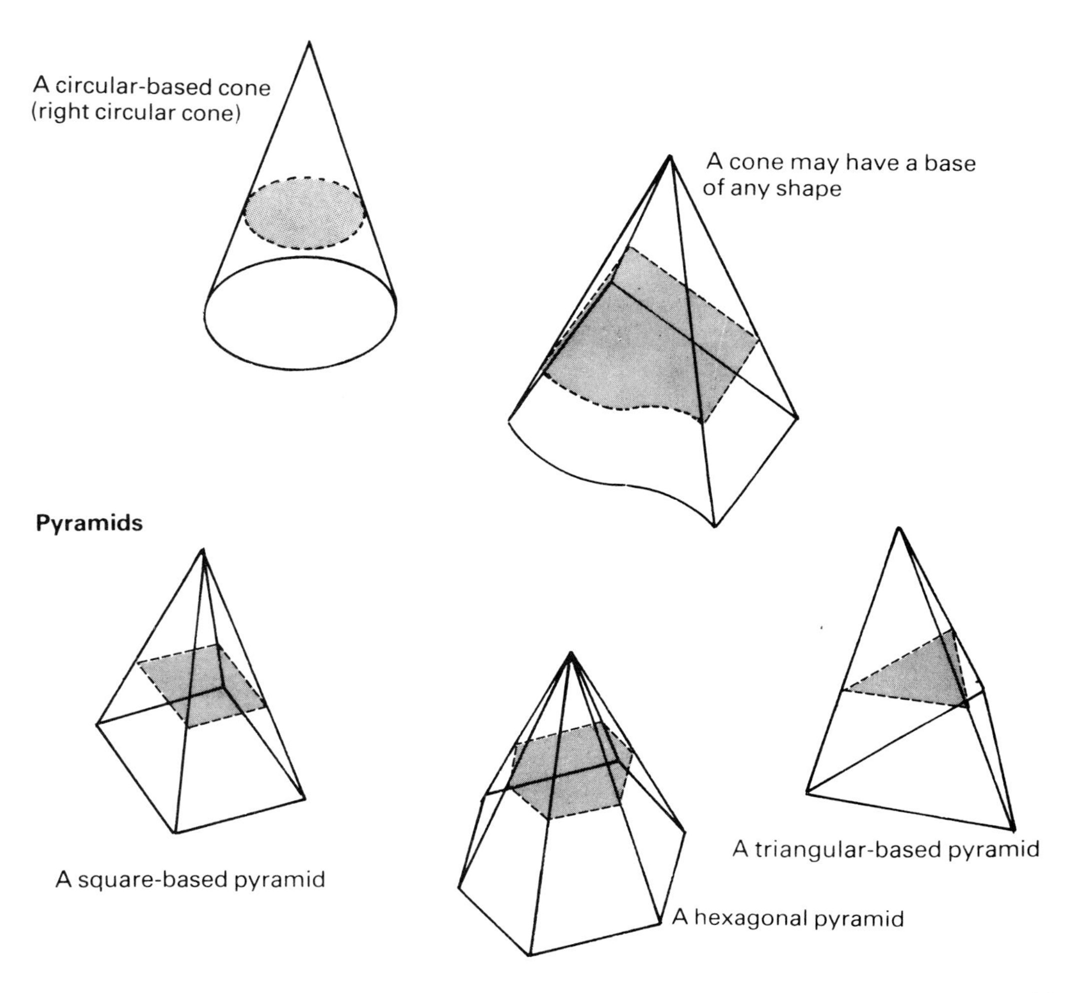

In a pyramid the base is a polygon and all edges from the base lead to one vertex (the apex). Thus all faces other than the base are triangular.
The cross-section must be the same shape as (similar to) the base.

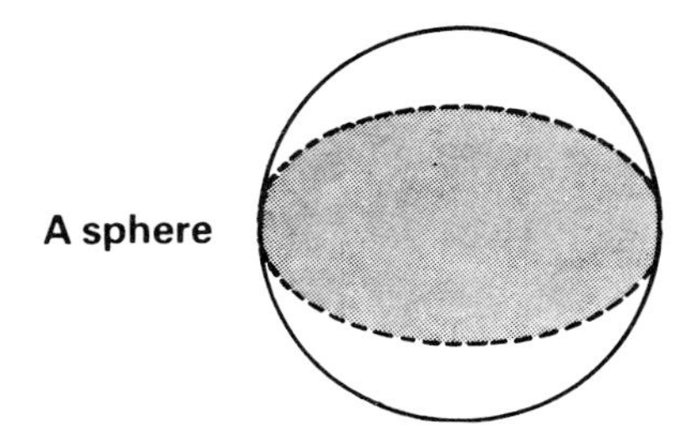

A sphere is a solid with an infinite number of planes of symmetry of which the shaded cross-section is just one.

1. In the cuboid:
   (a) How many corners or **vertices** are there?
   (b) How many **edges**?
   (c) How many surfaces or **faces**?
2. Repeat Question 1 for all the remaining prisms. Write your results in the following table:

Shape	No. of vertices	No. of faces	No. of edges
Cuboid			
Cube			
Triangular			
Trapezoidal			
Hexagonal			

   (a) What is true about all the faces other than the base and the top?
   (b) Is there any connection between the number of edges of the base and the total number of faces?
   (c) If the base had n edges how many faces would the prism have altogether?
   (d) Is there any connection between the number of vertices and the number of edges of the base?
   (e) Is there any connection between the total number of edges and the number of edges of the base?
   (f) Can you spot any connection between the number of vertices, faces and edges?
3. Try to draw neat sketches of all the prisms shown which have triangular or rectangular bases. If you have completed this you might like to try one of the prisms with more than four rectangular faces.
4. Draw a square of side about 10 cm on a piece of paper and then look at it from several different angles. Does it always look like a square? If you decide that it doesn't, is there any plane figure that it always looks like, no matter what angle you look at it from?
5. What is the difference between a cube and a cuboid?
6. What name would you give to the shape formed by a pile of 10 pence pieces placed one on top of the other?
7. If you wished to make a cube:
   (a) How many square faces would you need?
   (b) How would they be joined if you wanted them to be cut from one piece of cardboard?
   (c) Is your idea of how they have to be joined together the only way? Discuss this with the rest of the class and with your teacher.

   When this has been sorted out draw a plan of your cube (opened out) on a sheet of cardboard or paper and cut it out, without allowing any tabs to fix the unattached edges together. Then try to form it into a cube. This plan of your cube opened out is called a **net**. Try to make up the nets for some of the other solid shapes and then see if they work.

8. Study the nets given below and then answer the questions on the next page.

(a)

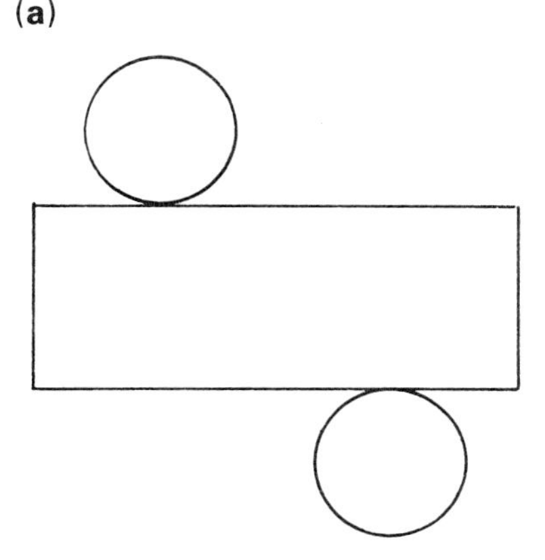

(b)

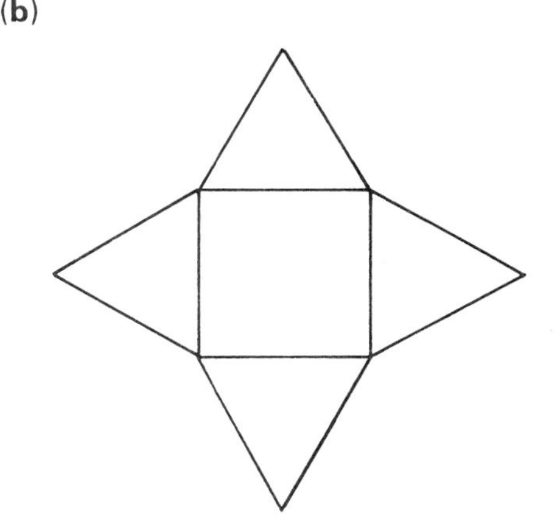

(c)

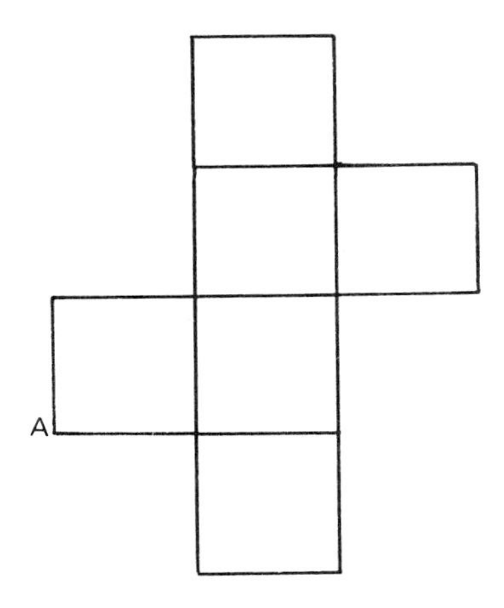

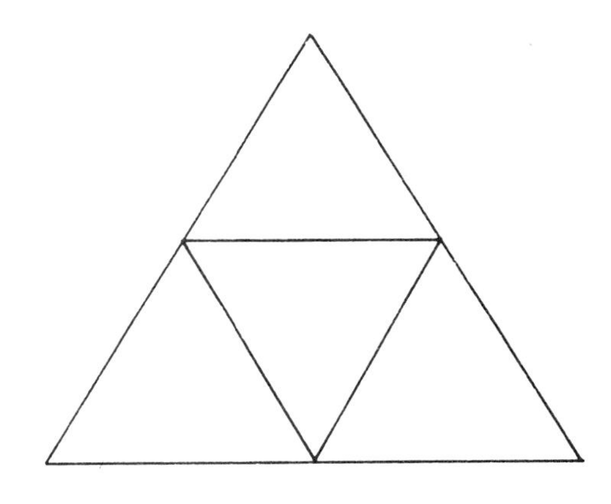

(e)

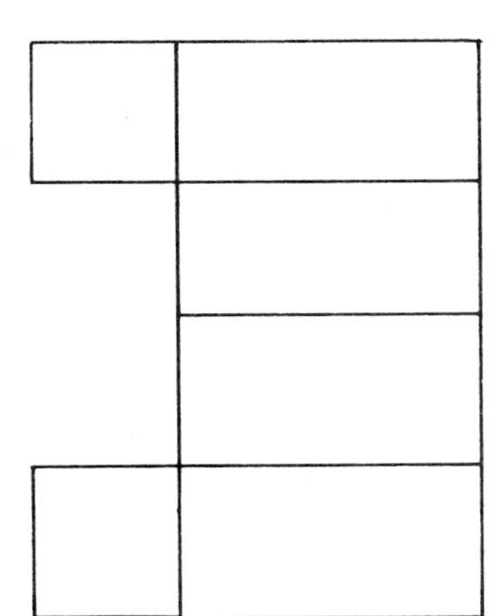

(f)

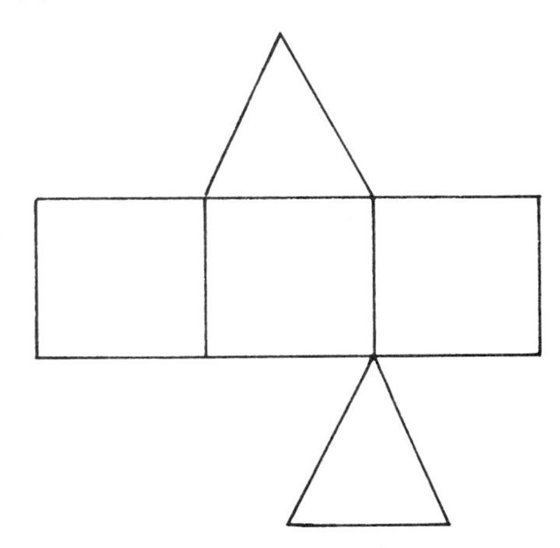

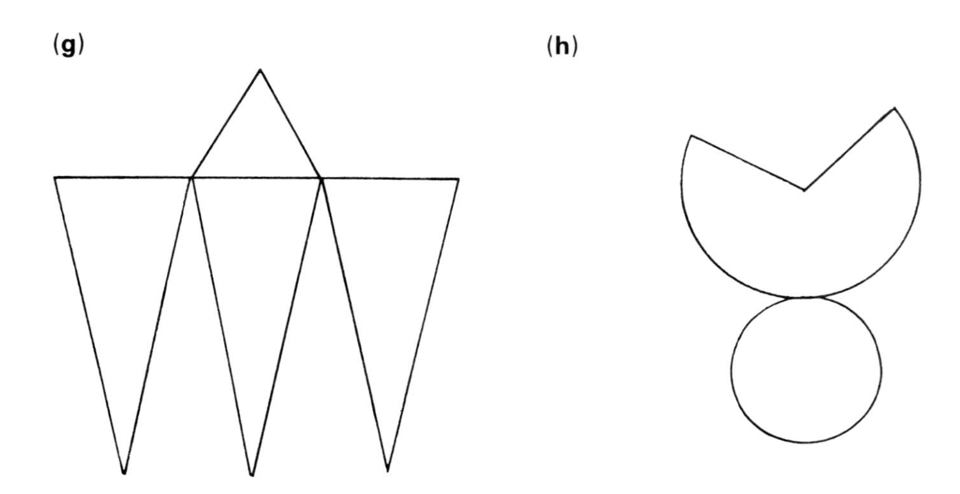

If you find it helpful, by all means copy these nets and cut them out carefully. We have not drawn any tabs to connect the edges, in order not to confuse you.

(i) In each case, give the name of the solid of which this shape is the net.

(ii) One of these nets is incorrectly drawn and two of them could be more sensibly drawn. Say which one is incorrect, giving reasons for your choice, and re-draw it correctly. Now say which two could be better drawn and re-draw them as you think they should have appeared.

(iii) In figure (c) make a copy of the net. Mark on it all the points which will meet at A when the shape is made up. (Think carefully how many edges meet at a vertex in this shape.)

28

Scale drawing

Exercise 28.1 *Scales*

The scale is 1 cm to 1 km.

1. How many km are represented by
(a) 3 cm **(b)** 4.5 cm **(c)** 11.6 cm?
(d) How many metres are represented by 0.5 cm?

2. How many cm would represent:
(a) 7 km **(b)** 9.4 km **(c)** 11.8 km **(d)** 700 m?

The scale is 1 cm : 10 m.

3. What distance is represented by
(a) 5 cm **(b)** 6.4 cm **(c)** 14 cm **(d)** 0.5 cm?

4. How many centimetres represent
(a) 60 m **(b)** 48 m **(c)** 120 m
(d) 115 m **(e)** 105 m **(f)** 7 m?

The scale is 1 cm : 100 km.

5. What distance is represented by
(a) 2 cm **(b)** 3.5 cm **(c)** 2.75 cm
(d) 12 cm **(e)** 11.4 cm **(f)** 0.5 cm?

6. What will these distances be represented by?
(a) 300 km **(b)** 650 km **(c)** 825 km
(d) 1250 km **(e)** 10 km

The scale is 1 cm : 2 m.

7. What distance is represented by
(a) 3 cm **(b)** 6.5 cm **(c)** 4.2 cm
(d) 9.7 cm **(e)** 11.25 cm?

8. What will be the scale measurements of
(a) 8 m **(b)** 15 m **(c)** 7.8 m
(d) 3.5 m **(e)** 0.5 m?

The scale is 1 cm : 50 m.

9. What distance is represented by
(a) 6 cm **(b)** 4.5 cm **(c)** 15 cm
(d) 1.8 cm **(e)** 0.1 cm?

10. What will be the scale measurements of
(a) 200 m **(b)** 375 m **(c)** 1250 m
(d) 630 m **(e)** 15 m?

The scale is 1 cm : 5 km.

11. What measurements correspond to

(**a**) 3 cm (**b**) 25 km (**c**) 4.5 cm
(**d**) 17 km (**e**) 0.5 cm (**f**) 4 km?

Exercise 28.2 *Scale drawings*

1. A football field is 100 m long by 60 m wide. By scale drawing, find the diagonal distance across the ground.
2. A picture which measures 40 cm by 30 cm is surrounded by a mount 10 cm wide. Draw a diagram to illustrate this.
3. A rectangular courtyard measures 10 metres by 7 metres. Exactly in the centre there is a circular fishpond which has a radius of 2 metres. Draw a plan of the courtyard.
4. The top of a ladder reaches 6 m up the side of a house when its foot is 1.5 m from the wall. How long is the ladder?
5. A kite is flying on the end of a string 30 m long. The string makes an angle of 57° with the ground. How high is the kite?
6. A flagpole is 10 m high. A guy rope from the top of the pole is fastened to the ground at a point 5.5 m from the base. How long is the rope?
7. At two points on opposite sides of a tree the angles of elevation of its top are 37° and 49°. If the distance between the points is 85 m, how high is the tree?
8. From a point 55 m from the base of a vertical tower the angle of elevation is 30°. How high is the tower?
9. The top of an aqueduct is 60 m above ground level. The angle of depression to a point in the valley below is 23°. How far away is the point from the base of the aqueduct?
10. A bookcase is 2 m long and 2.5 m high. The top 4 shelves are 25 cm apart, while the bottom part is taken up with 3 shelves evenly spaced, with the floor level as the bottom shelf. Draw a front view of this bookcase. (Disregard the thickness of the wood.)

Exercise 28.3 *Bearings – drawing*

From a point P draw lines to represent the following bearings:

1. Due West
2. 047°
3. 118°
4. 147°
5. 190°
6. 210°
7. 300°
8. North-West
9. 127°
10. 335°

Exercise 28.4 *Bearings – measurement*

In each case, what is the bearing of (**a**) A from B
(**b**) B from A?

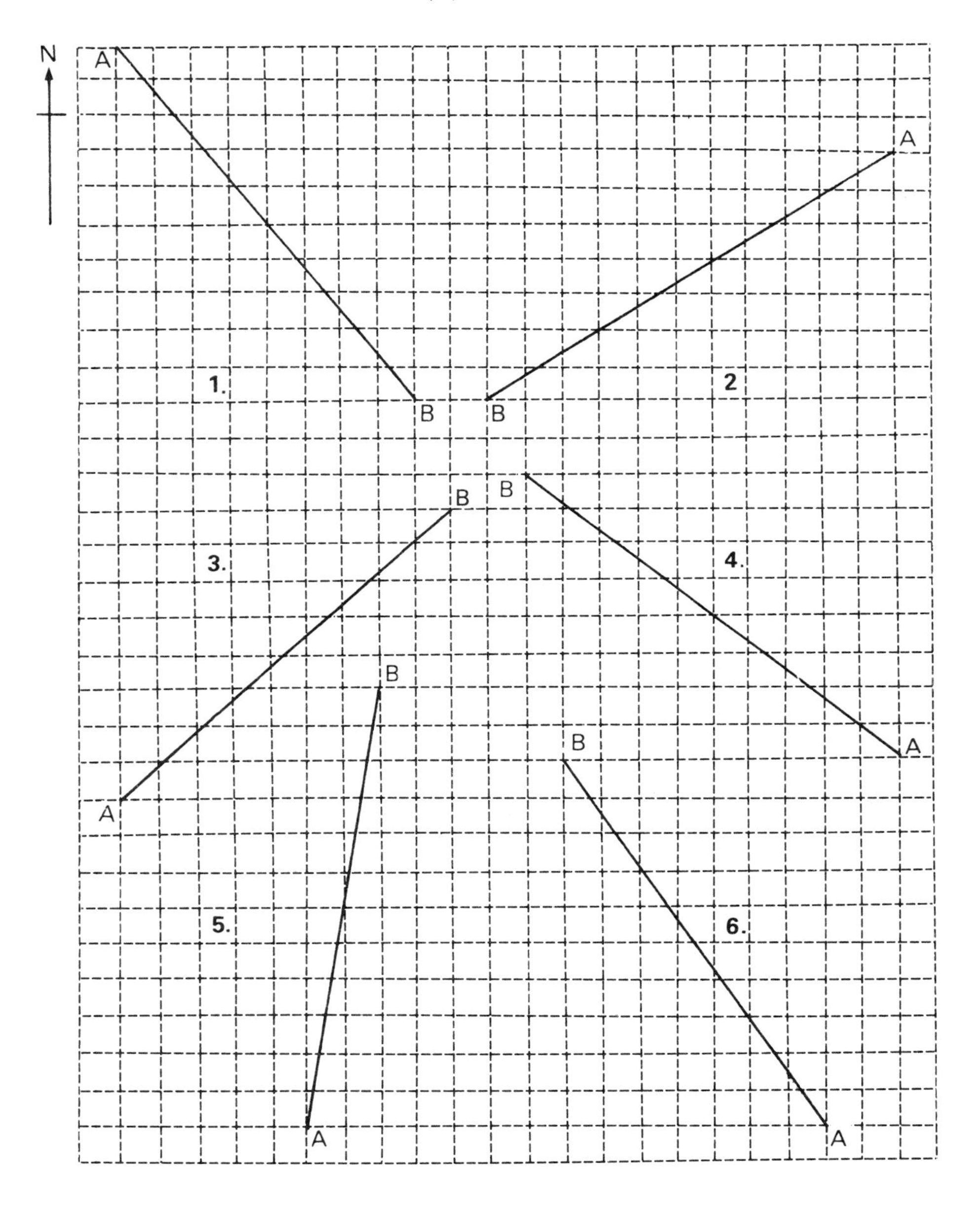

Exercise 28.5 *Scale drawings with bearings*

1. B is 5 km East of A, C is 6 km North of B and D is 2 km West of C. What is
   (**a**) the distance of A from D
   (**b**) the bearing of A from D
   (**c**) the bearing of D from A?
2. Q is 70 km N.E. of P while R is 40 km S.E. of Q. What is the distance and bearing of P from R?
3. X is 20 km West of Y and 9 km North of Z. What is the distance and bearing of Z from Y?
4. The church is 12 km from the manor on a bearing of 225° and the station is 8 km from the church on a bearing of 315°. What is the distance and bearing of the manor from the station? How can 225° and 315° be written using points of the compass?
5. A liner is 9.6 km North of a buoy while the lighthouse is 8 km from the ship on a bearing of 053°. How far is the buoy from the lighthouse and on what bearing?
6. A pilot flies due East to an aerodrome 860 km away, but seeing that it is in enemy hands he turns North West and flies 750 km to another aerodrome. How far is he from where he started and what is his bearing from there?
7. A yacht sails 7.6 km on a bearing of 270° to a buoy where it changes course and sails on a bearing of 135° for 5.9 km. It then sails back to the starting point. How long is the total course sailed? On what bearing does it sail on the last leg?
8. From where she sits Barbara sees a windmill 750 m away on a bearing of 320° and a telegraph pole 580 m away on a bearing of 060°. How far is the windmill from the telegraph pole?
9. Ryde is 16 km from Ventnor on a bearing of 011° and Yarmouth is 24 km from Ryde on a bearing of 263°. How far is Yarmouth from Ventnor and on what bearing?
10. From an observation post an enemy foxhole is spotted 650 metres away on a bearing of 040°. The maximum range of enemy weapons is 400 m. Between what bearings is the sector to avoid if an advance is to pass the enemy in safety?
11. A and B are points on a path which runs North–South and A is 500 m due North of B. An oak tree is on a bearing of 130° from A and 070° from B. How far is it from A to B via the oak tree?
12. Two coastguard stations, P and Q, are 7 km from each other with P being due East of Q. A distress flare is seen on a bearing of 160° from Q and 206° from P. There is a lifeboat stationed 2 km due East of P. On what bearing must the lifeboat set out to reach the point where the flare was let off?
13. On Tutti Island there is a very tall palm tree which is 450 m from a mooring post on a bearing of 330°. It is known that a bag of piastres is buried 275 m from the palm tree, equidistant between it and the mooring post and to the West of the post. On what bearing should one set out from the mooring post, hoping to find the treasure?
14. A ship sails at 20 km/h for $\frac{3}{4}$ of an hour on a bearing of 040°. It then changes course to a bearing of 335° and increases speed to 30 km/h. How long is it before the ship is due North of its original position?
15. Clare is at C and can walk at 8 km/h while Bill is at B and can walk at 6 km/h. At the same time they start walking towards A which is 7.5 km due West of B. B is 6 km from C on a bearing of 190°. Who arrives at A first?

29

Loci

Exercise 29.1 *Simple loci*

1. Construct the locus of a point which is always 4 cm from a given point A.
2. A and B are two points 8 cm apart. Construct the set of points that are equidistant from A and B.
3. AB = 8 cm. Construct the set of points that are equidistant from A and B.
4. AB = 8 cm. Construct the set of points that are 2.5 cm from AB.
5. AB and CD are parallel lines 5 cm apart. Construct a figure to show the locus of points equidistant from the two lines.
6. $\widehat{BAC} = 70°$. Construct the locus of points equidistant from AB and AC.
7. AB and CD are two lines that intersect in the shape of an X. Construct the locus of points equidistant from AB and CD.

Exercise 29.2 *Further loci*

1. AB = 4 cm. Construct the locus of a point C such that AC = 3 cm and of a point D such that BD = 2 cm. Mark in the points that are both 3 cm from A and 2 cm from B.
2. Draw a line 6 cm long and letter it AB. Then draw the locus of a point which moves at a distance of 4 cm from A and of a point that is always 3.5 cm from B. If these loci intersect at X and Y measure $\widehat{AXB}$ and $\widehat{AYB}$.
3. Draw AB = 6 cm. Construct the set of points that are 2 cm from B and the set of points that are 3 cm from A. How many times do these loci intersect?

For questions 4–7 use copies of the figure below, drawn accurately (it is not drawn accurately here).

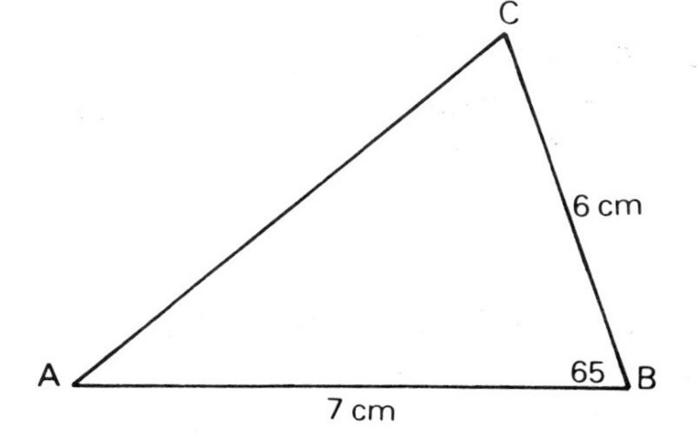

4. Construct
   (**a**) the locus of points 2 cm from BC
   (**b**) the locus of points 2.5 cm from C.
5. Construct
   (**a**) the locus of points equidistant from B and C
   (**b**) the locus of points equidistant from A and B.

   Does the point where they meet have any importance to the triangle? If so, explain what it is.

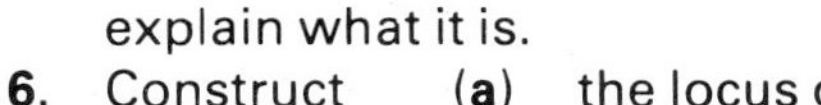

6. Construct (**a**) the locus of points equidistant from AB and BC
   (**b**) the locus of points equidistant from AB and AC.

Does the point where they meet have any importance to the triangle? If so, explain what it is.

7. Construct (**a**) the locus of points equidistant from A and C
   (**b**) the locus of points equidistant from CA and CB.

   Let them meet at O: measure CO.

8. ABCD is a square of side 5 cm. Construct the locus of points equidistant from AD and BC and the set of points equidistant from AB and AD.
9. WXYZ is a rectangle in which WX = 6 cm and XY = 4 cm. Construct the locus of points equidistant from WX and YZ and the set of points equidistant from WX and XY.
10. On a grid the locus of a point X is such that $x = 4$ and the locus of a point Y is such that $y = -2$. What are the coordinates of the intersection of these loci?

Exercise 29.3 *Two-dimensional loci*

1. AB is 6 cm long. Shade the set of points that are less than 4 cm from AB and more than 3 cm from A.

For Questions 2–5 use copies of the figure below, drawn accurately (it is not drawn accurately here).

2. Shade the points inside the triangle that are less than 4 cm from PQ and nearer to RQ than RP.
3. Shade the points inside the triangle that are more than 3 cm from P and nearer to Q than R.
4. Shade the points inside the triangle that are nearer to P than Q and nearer to Q than R.
5. Shade the points inside the triangle that are nearer to PQ than PR and nearer to QR than QP.

For Questions 6 and 7 use a 4 cm square, lettered as in the figure:

6. Shade the points inside the square that are nearer to AB than CD and nearer to BC than AD.
7. Shade the points inside the square that are nearer to AB than AD and nearer to BC than BA.

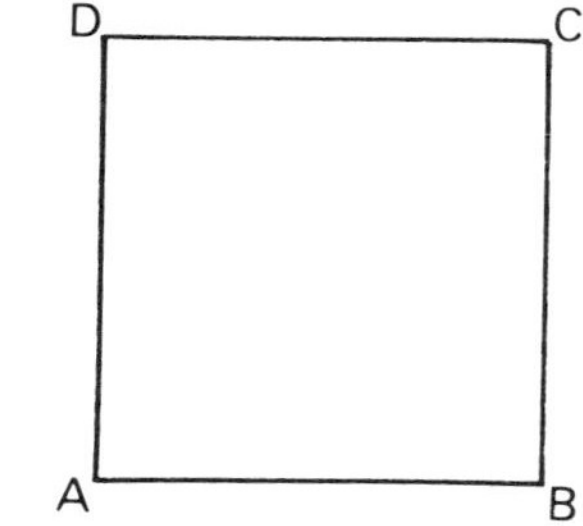

For questions 8 and 9 use this rectangle:

8. Shade the points inside the rectangle that are nearer to AB than CD and nearer to BC than AD.
9. Shade the points inside the rectangle that are nearer to AB than AD and nearer to AD than DC.

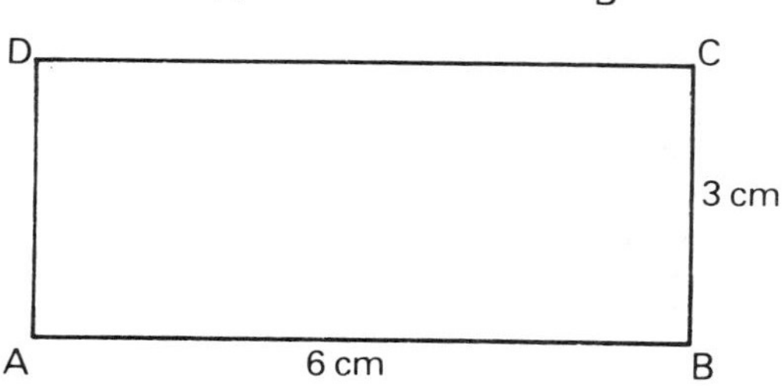

10. AB = 6 cm
X = {points less than 4 cm from A}
Y = {points more than 3 cm from B}
Shade $X \cap Y$.

11. In the square WXYZ which has side 4 cm in length:
A = {points nearer to X than Z}
B = {points that are less than 4 cm from W}
Shade $A \cup B$.

12. Draw rectangle ABCD in which AB = 7 cm and BC = 4 cm.
P = {points inside the rectangle nearer to B than D}
Q = {points inside the rectangle nearer to AB than BC}
Shade $P \cap Q$.

30

Further transformations

Exercise 30.1 *Translations*

The figure on the right shows the triangle PQR under a translation.

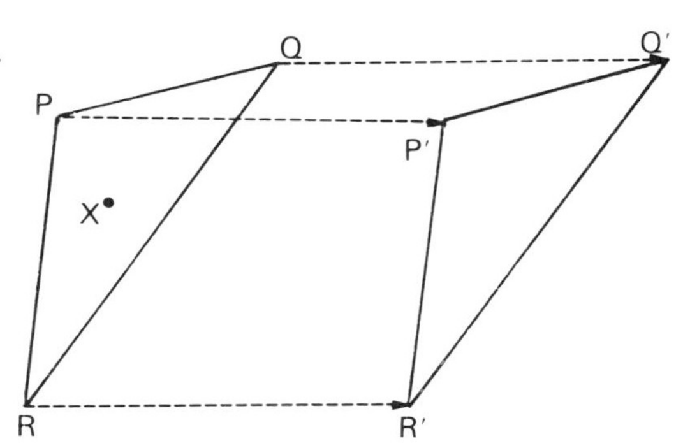

1. Measure the lengths of PP′, QQ′, RR′. What do you notice?
2. What do you notice about the lines PP′, QQ′, RR′ in terms of their direction?
3. If **PP**′ is described as a vector having length and direction, give two other vectors equivalent to **PP**′.
4. Use this to find the image of point X under this translation.

The figure on the right shows the triangle ABC under a translation **AA**′ and a further translation of **A**′**A**″.

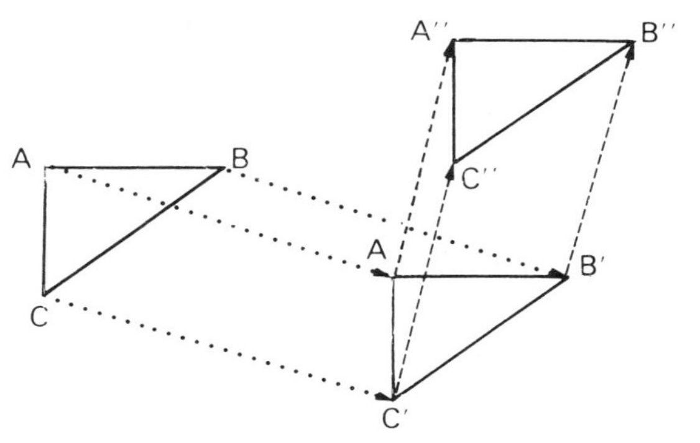

5. Give two vectors equivalent to **AA**′.
6. Give two vectors equivalent to **A**′**A**″.
7. Measure the lengths of AA″, BB″ and CC″. What do you notice?
8. What do you notice about the directions of the lines AA″, BB″, CC″?
9. Could **AA**″, **BB**″ and **CC**″ be described as equivalent vectors?
10. Is it possible to map triangle ABC onto triangle A″B″C″ by a single translation? Give a reason.

In the figure on the right, triangle PQR is translated by a vector **QQ**′.

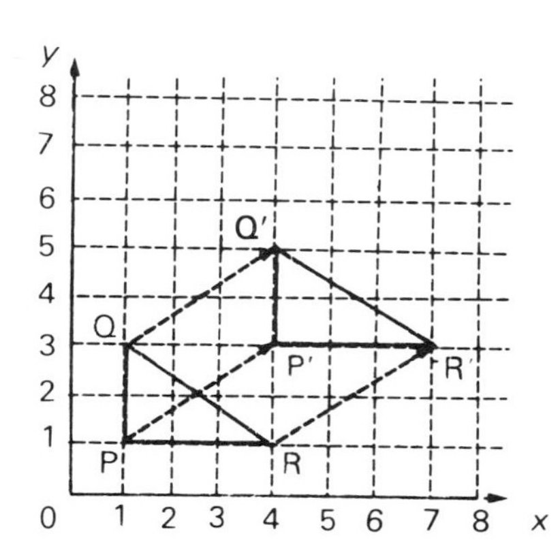

11. How far in a direction parallel to the x-axis has Q moved to map onto Q′?
12. How far in a direction parallel to the y-axis has Q moved to map onto Q′?
13. The vector **QQ**′ is described by the column vector $\begin{pmatrix}3\\2\end{pmatrix}$. What do you think the 3 and 2 describe?

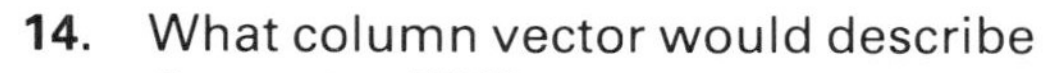

14. What column vector would describe the vector **RR**′?

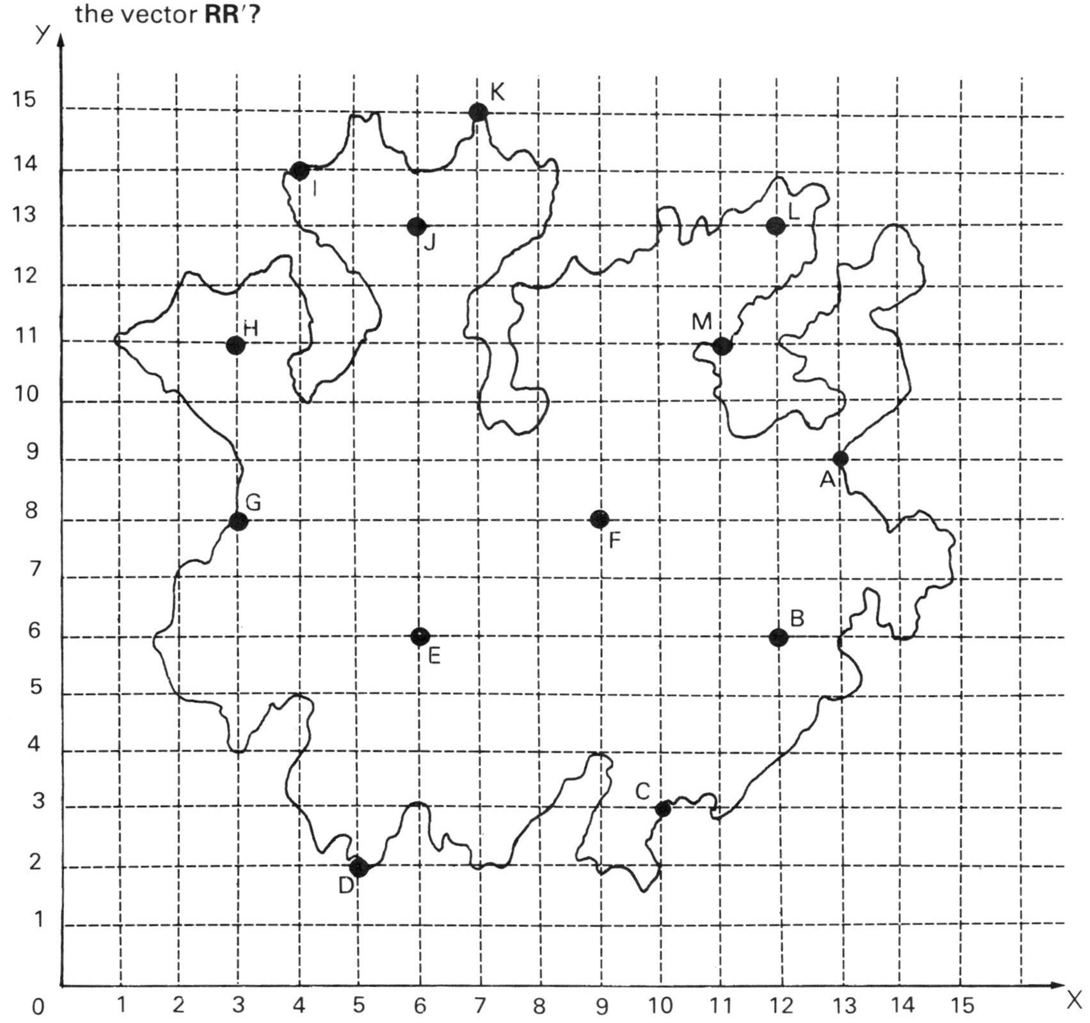

List of names

A – Albertville
B – Brenda's Beacon
C – Clive's Cove
D – Dianatown
E – Enryston
F – Frieda's Folly
G – Greg's Gorge
H – Helen Heights
I – Ian's Isle
J – Jennystone
K – Keith's Kop
L – Linda's Leap
M – Martin's Mill

From the figure above give the column vectors to describe these movements:

	From	To		From	To
15.	Enryston	Frieda's Folly	**19.**	Greg's Gorge	Helen Heights
16.	Clive's Cove	Dianatown	**20.**	Martin's Mill	Brenda's Beacon
17.	Greg's Gorge	Jennystone	**21.**	Keith's Kop	Enryston
18.	Frieda's Folly	Greg's Gorge	**22.**	Ian's Isle	Martin's Mill

23. Describe a route around the coast by sea to get to Keith's Kop from Albertville. (Use a series of column vectors.)
24. Describe an overland route to get to Linda's Leap from Helen Heights via Frieda's Folly. (Use a series of column vectors to do this.)
25. Describe, using column vectors, a route around the island from Dianatown. What do you notice about the sum of these column vectors?
26. Give the column vector from Greg's Gorge to Enryston and that from Frieda's Folly to Enryston. What does this suggest about the distance from Greg's Gorge to Enryston and from Enryston to Frieda's Folly?

Give the column vectors required to map the first of each pair of points onto the second.

27. $(0,0) \rightarrow (2,5)$
28. $(1,2) \rightarrow (3,5)$
29. $(0,0) \rightarrow (-1,-2)$
30. $(4,2) \rightarrow (3,0)$
31. $(-2,1) \rightarrow (2,3)$
32. $(1,4) \rightarrow (-3,0)$
33. $(1,3) \rightarrow (-2,1)$
34. $(5,0) \rightarrow (-2,4)$
35. $(1,1) \rightarrow (-3,-1)$
36. $(a,b) \rightarrow (x,y)$

In the figure on the right:

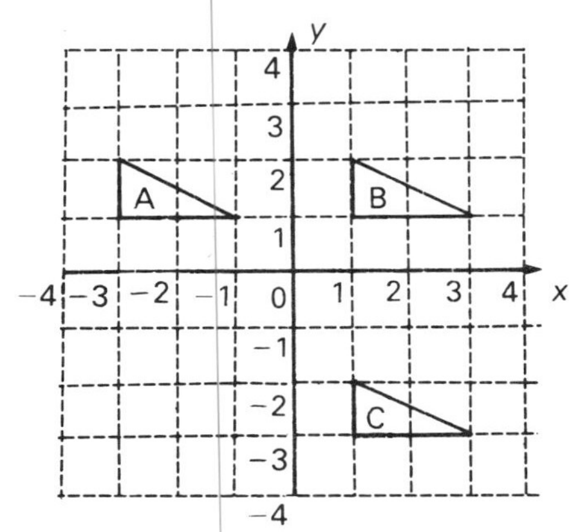

37. Give the column vector describing the translation of triangle A onto triangle B.
38. Give the column vector describing the translation of triangle B onto triangle C.
39. Give a single vector that would describe the translation of triangle A onto triangle C.
40. What do you notice about the sum of your answers to Questions 37 and 38 and your answer to Question 39?
41. What single translation would take triangle C onto triangle A?
42. How does this compare to your answer to Question 39?

Give the images of the following after they have been translated by the vector $\begin{pmatrix} 3 \\ -2 \end{pmatrix}$:

43. $(0,0)$
44. $(1,2)$
45. $(3,4)$
46. $(-1,1)$
47. $(4,-2)$
48. $(6,2)$
49. $(-3,4)$
50. $(-3,-4)$
51. $(4,-3)$
52. $(-3,2)$
53. $(-4,0)$
54. (x,y)

In the figure on the right:

55. What column vector would describe the translation of triangle A onto triangle B?

56. What column vector would describe the translation of triangle B onto triangle C?

57. What column vector would describe the translation of triangle A onto triangle C?

58. Describe the vector required to map triangle C onto triangle A.

59. How does this compare with your vector in Question 57?

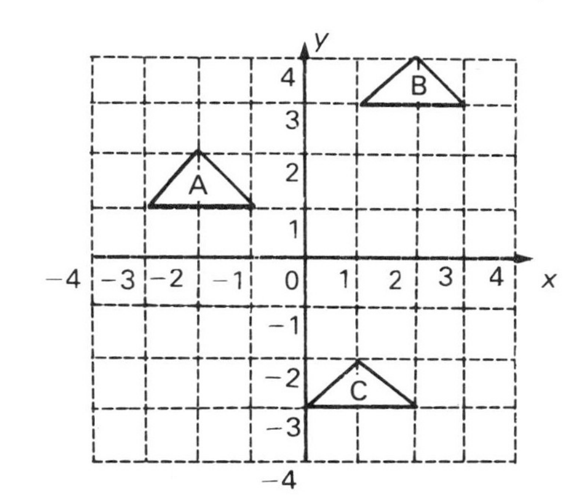

The points A(0,0), B(3,0), C(x,y) and D(1,2) form a parallelogram.

60. Give the coordinates of point C if both x and y are positive.

61. What is the column vector **AD**?

62. Which other side of the parallelogram should have a column vector equivalent to vector **AD**?

63. Use this idea to check your answer to Question 60.

64. Now use this idea to find another point for the vertex C but this time with the x value negative and the parallelogram lettered ABDC.

65. Repeat this for a third position of C such that x is positive but y is negative.

If $\mathbf{v}=\begin{pmatrix}2\\3\end{pmatrix}$, $\mathbf{w}=\begin{pmatrix}0\\3\end{pmatrix}$, $\mathbf{x}=\begin{pmatrix}2\\0\end{pmatrix}$ and $\mathbf{z}=\begin{pmatrix}-2\\-3\end{pmatrix}$, what is the final column vector equivalent to each of these?

66. $\mathbf{w}+\mathbf{x}$

67. $\mathbf{v}+\mathbf{w}$

68. $\mathbf{w}+\mathbf{x}+\mathbf{z}$

69. $2\mathbf{v}$

70. $3\mathbf{w}$

71. $2\mathbf{v}+3\mathbf{w}$

72. $2\mathbf{w}+2\mathbf{x}+2\mathbf{z}$

73. $\mathbf{v}-\mathbf{w}$

74. $\mathbf{w}-\mathbf{x}$

75. $\mathbf{v}-\mathbf{w}-\mathbf{x}$

Exercise 30.2 *Mixed transformations*

Question 1–24 refer to the figure on the right.

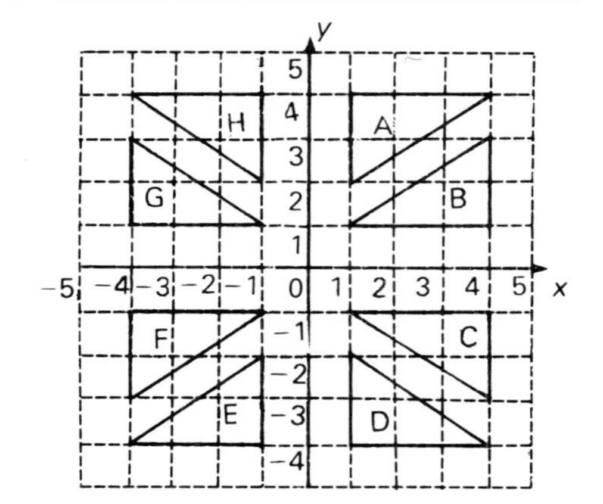

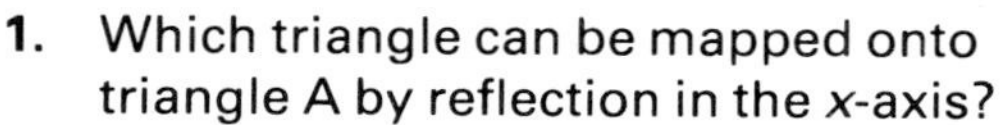

1. Which triangle can be mapped onto triangle A by reflection in the x-axis?

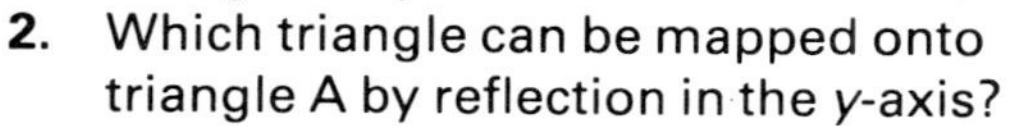

2. Which triangle can be mapped onto triangle A by reflection in the y-axis?

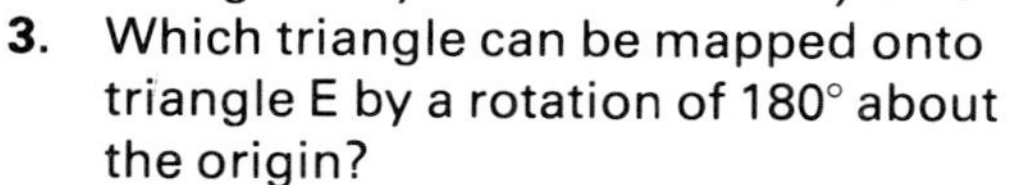

3. Which triangle can be mapped onto triangle E by a rotation of 180° about the origin?
4. Which triangle can be mapped onto triangle G by a rotation of 180° about the point (−2.5,2.5)?
5. Which single transformation would map triangle F onto triangle B?
6. Which single transformation would map triangle C onto triangle B?
7. Which single transformation would map triangle E onto triangle H?
8. Which single transformation would map triangle H onto triangle C?
9. Which single transformation would map triangle D onto triangle G?
10. Name two reflections that would combine to map triangle B onto triangle F.
11. Name two rotations that would map triangle D onto triangle G if applied one after the other.
12. Can triangle A be reflected by a single reflection onto triangle B?
13. Give the mirror lines of two reflections that would map triangle F onto triangle E if combined.
14. Give two transformations that would combine to map triangle A onto triangle D.

R is a rotation of 180° about the point (0,0), X is a reflection in the x-axis, Y is a reflection in the y-axis and T is a translation by the vector $\begin{pmatrix}5\\5\end{pmatrix}$. Remember that XY(△A) means 'Y followed by X', perhaps better explained by X(Y(△A)). Complete the following statements:

15. R(△A) = △
16. T(△F) = △
17. X(△C) = △
18. Y(△G) = △
19. XY(△A) = △
20. X(△) = △F
21. Y(△) = △E
22. XY(△) = △H
23. TR(△) = △B
24. RT(△) = △F

Questions 25–32 refer to the figure on the right.

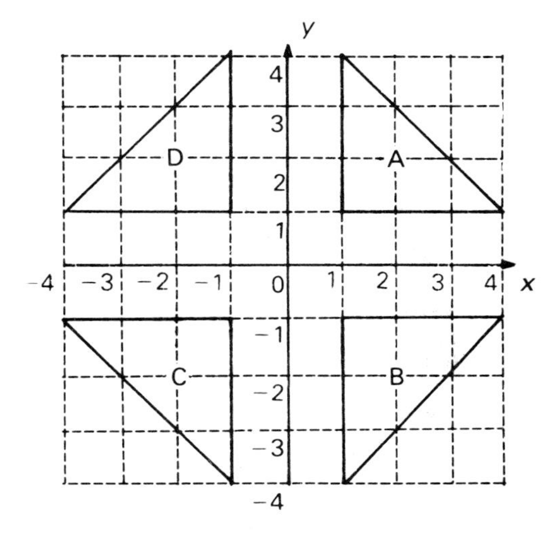

25. Is it possible to say from the figure if triangles A, B, C and D are similar or congruent? Could they be both? Give reasons for your answer.
26. In which line could triangle A be reflected onto triangle D?
27. Through what angle and about which point could A be rotated onto B?
28. In which mirror line could triangle A be reflected onto triangle C?
29. What single transformation could map triangle B onto triangle A?
30. Could triangle A be translated onto any of the other triangles? Give a reason for your answer.
31. How many single transformations could be used to map triangle A onto triangle D?
32. Describe the symmetry of the figure made up of the axes and triangles.

Questions 33–40 refer to the figure on the right.

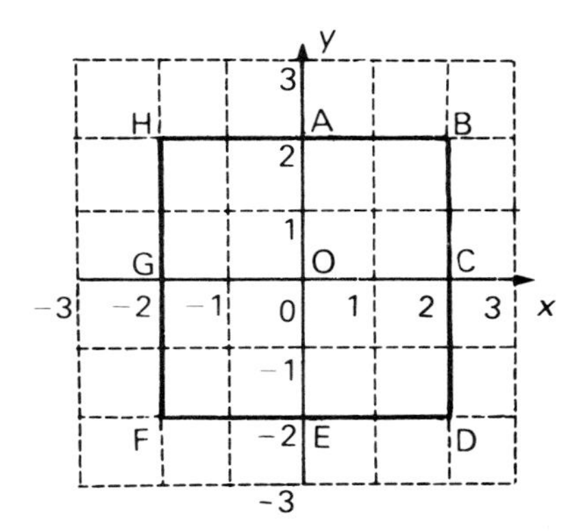

33. What transformation would map ABCO onto EFGO?
34. What transformation would map ABCO onto GHAO?
35. What transformation would map ABCO onto GOEF?
36. In which mirror line could ABCO be reflected onto GFEO?
37. In which mirror line could CDEO be reflected onto GFEO?
38. In which mirror line could GFDC be mapped onto GHBC by reflection?
39. In which mirror line could ABCO be reflected onto AOCB?
40. Where would be the centre of enlargement and what would be the scale factor to enlarge ABCO onto HBDF?

Questions 41–45 refer to the figure on the right.

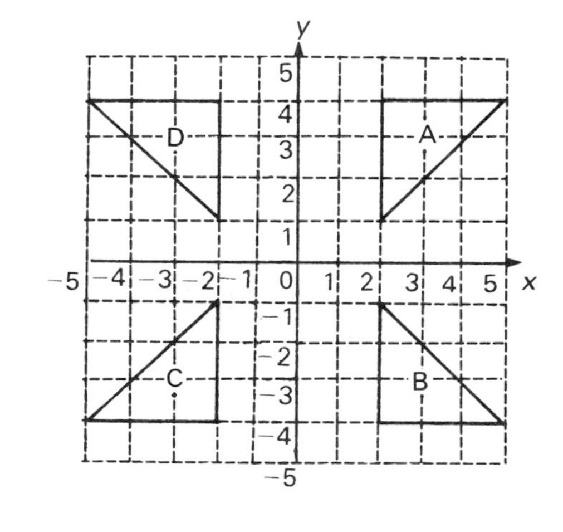

41. What single transformation would map △A onto △B?

42. What single transformation would map △B onto △D?

43. Give the centre and angle of rotation that would map △A onto △C.

44. About which point could △B be rotated onto △C?

45. How would you find the angle of rotation for the rotation of triangle B onto triangle C?

In the figure on the right all three triangles are equilateral.

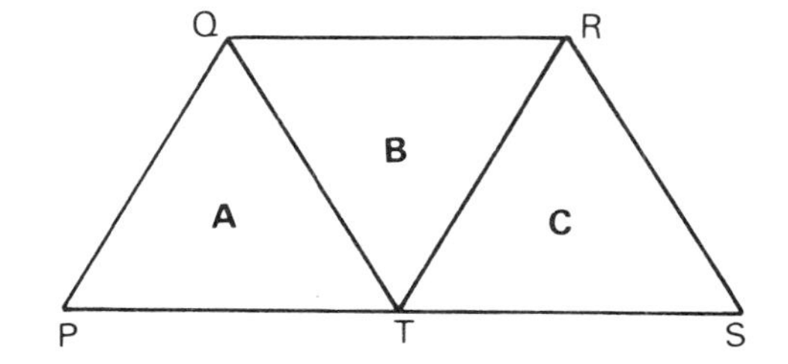

46. In how many different ways can triangle A be mapped onto triangle B? (Give all the details.)

47. In how many different ways can triangle A be mapped onto triangle C? (Once again give all details.)

Questions 48–55 refer to the figure on the right.

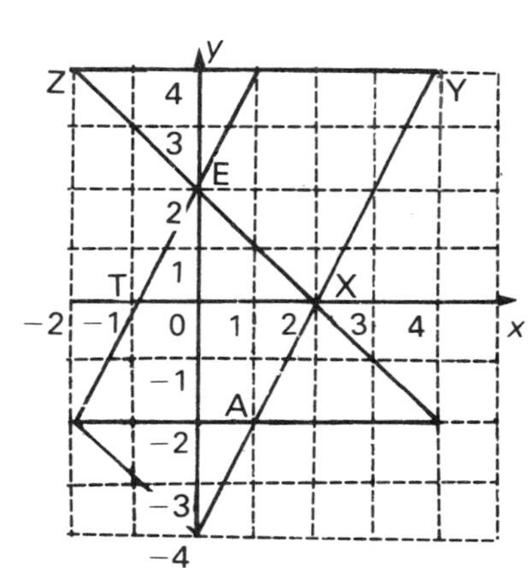

48. What single transformation would map triangle ABX onto triangle ADC?

49. What single transformation would map triangle ABX onto triangle TXE?

50. What centre of enlargement and scale factor would map triangle ABX onto triangle DBE?

51. What single transformation would map triangle XYZ onto triangle EDB?

52. What single transformation would map triangle DAC onto triangle ZFE?

53. What single transformation would map triangle DAC onto triangle ZYX?

54. Give the single transformation that would map triangle XAB onto triangle EFZ.

55. What single transformation would map triangle ZYX onto triangle ZFE?

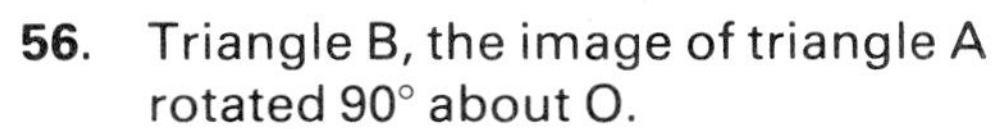

Make a copy of the figure on the right and then draw in the following:

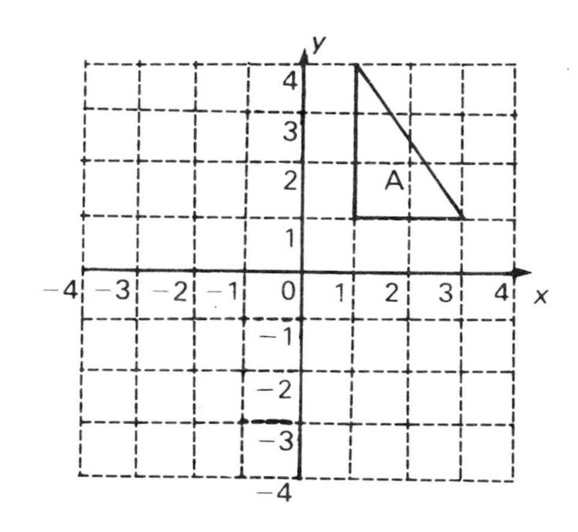

56. Triangle B, the image of triangle A rotated 90° about O.

57. Triangle C, the image of triangle A rotated 180° about O.

58. Triangle D, the triangle that completes the rotational symmetry of the figure.

59. What is the order of rotational symmetry in this case?

In the figure on the right, find in each case the single transformation that will map:

60. ABCD onto EFGH

61. ABCD onto GHEF

62. WDXY onto RSPQ

63. PQRS onto ABCD

64. EFGH onto PQRS.

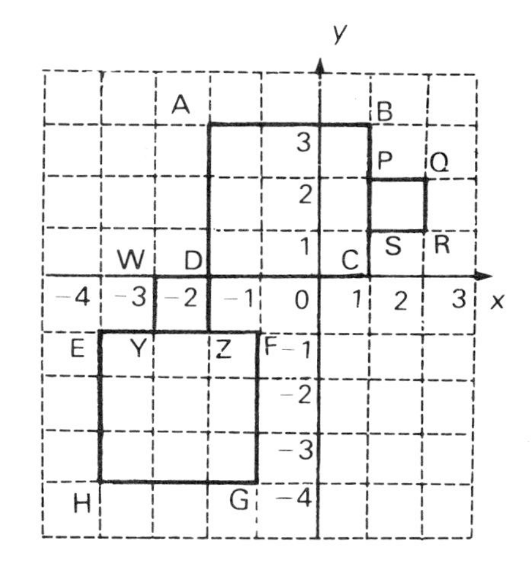

Make a copy of the figure on the right.

65. In how many ways can square A be transformed onto square B? (Give details.)

66. In how many ways can square B be mapped onto square C? (Give details.)

67. What transformation would map the whole figure onto itself, with D mapping onto A and C onto B?

68. Comment on the symmetry of the figure.

69. Plot the points A(1,1), B(3,1), C(1,3) and join them to form the triangle ABC.

70. Draw the image of triangle ABC under a reflection in the x-axis.

71. What are the coordinates of A′, B′ and C′, the images of A, B and C respectively?

72. Translate triangle ABC by a vector of $\begin{pmatrix} 0 \\ -5 \end{pmatrix}$. Mark the image A″B″C″.

73. What single transformation would map A′B′C′ onto A″B″C″?

Look at the figure opposite. In each case, give a single transformation that will map:

74. ABCM onto DEFM
75. IMGH onto ABCM
76. MDEF onto MJKL
77. KLMJ onto MABC.
78. What kinds of symmetry has the figure?

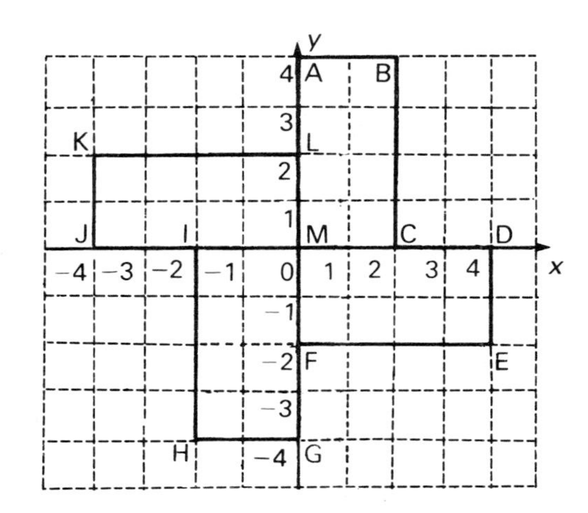

Questions 79–88 refer to the figure opposite.

79. Name three triangles that can be mapped onto triangle AHG by rotation about O.
80. What single transformation would map triangle BIC onto triangle FLE?
81. What single transformation would map triangle OBC onto triangle OEF?
82. What single transformation would map triangle OBC onto triangle OFE?
83. What name is given to the figure GHIJKL?

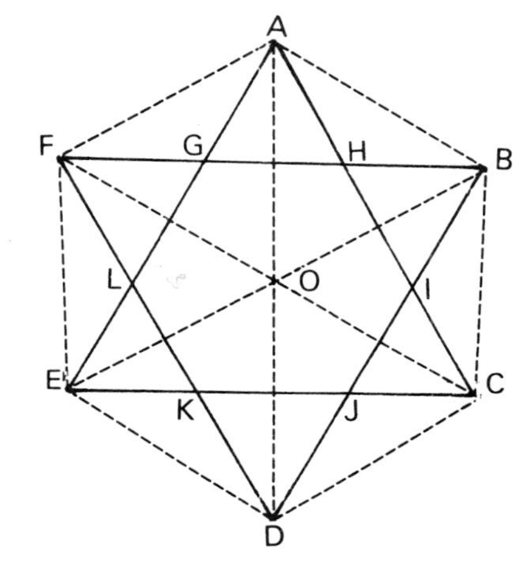

84. In what mirror line could triangle AFB be mapped onto triangle DEC?
85. Name the centre and angle of rotation required to rotate triangle ABE onto triangle DEB?
86. What single transformation would map triangle FOE onto triangle BCD?
87. What name is given to the figure GBJE?
88. Give two transformations which when applied consecutively would map GHIJKL onto ABCDEF.